An Open-Source Research Platform for Heterogeneous Systems on Chip

Andreas Dominic Kurth

An Open-Source Research Platform for Heterogeneous Systems on Chip

Hartung-Gorre Verlag Konstanz
2022

Reprint of Diss. ETH No. 28249

SERIES IN MICROELECTRONICS **VOLUME 244**

edited by **Luca Benini**
Qiuting Huang
Taekwang Jang
Mathieu Luisier
Christoph Studer
Hua Wang

Bibliographic information published by Die Deutsche Nationalbibliothek

Die Deutsche Nationalbibliothek lists this publication in the Deutsche Nationalbibliografie; detailed bibliographic data is available in the internet at http://dnb.dnb.de.

First edition 2022

HARTUNG-GORRE VERLAG KONSTANZ
http://www.hartung-gorre.de

ISSN 0936 - 5362
ISBN-10: 3-86628-774-7
ISBN-13: 978-3-86628-774-7

Acknowlegments

Completing a PhD was the most challenging but also the most interesting part of my education so far. Not only have I learned a lot technically and professionally, but I think I have also grown substantially as a person through this remarkable experience. For this, I am deeply grateful to the people who have supported me in reaching this goal.

I thank my supervisor and examiner, Luca Benini, for giving me the chance of pursuing this PhD project with him, for his highly experienced and critical feedback, and for establishing and managing a very productive research environment. I thank my co-examiner, Andrea Marongiu, and my co-referee, Davide Bertozzi, for their time and effort to review my thesis and provide valuable feedback. Thanks also go to the professional volunteers at organizations such as IEEE, ACM, and the RISC-V Foundation, especially conference contributors and journal reviewers, for forming a thriving global research community.

ETH Zurich in general and the Integrated Systems Laboratory (IIS) in particular were a highly productive and inspiring environment for me, and I am grateful to the persons who shaped ETHZ and IIS for me: Norbert and Hubert for excellent teaching and paving the road of IIS and DZ; Frank for maintaining the highly productive lab environment and managing research collaborations; Christoph, Hansjörg, Beat, and Christine for making our work much easier with their expertise and help. I have also benefited from collaborations with the University of Bologna and the work of many colleagues there, most notably Davide, Giuseppe, Francesco, Alessandro, and Germain. At ETHZ, I furthermore really enjoyed joint projects with the Scalable Parallel Computing Laboratory (SPCL), and I would like to thank Torsten, Salvatore, and Timo for broadening the scope of my knowledge and work to data centers and off-chip networks as well as Tobias for the refreshing collaborations around compilers and hardware construction.

My colleagues gave the research group a fun and inspiring work atmosphere, and special thanks go to: Pirmin for encouraging me to pursue a PhD, mentoring me through my first year and publications, and his fundamental contributions to the initial HERO platform; Pascal for being a role model in terms of engineering and productivity; Florian G. for fearlessly venturing into mixed-signal design and sharing his lessons learned; Michael S. for being inspiring as jack of all trades (and master of many, even in sports); Michael G. and David for laying the technological foundation for much of our work and forming our office culture; Björn for very fruitful collaborations in general and his contributions on compilers specifically; Fabian for ingenious and fun side projects (many of which turned out to have much more impact than we had imagined!); Florian Z. and Stefan for the chip frenzy and sharing the gained insights; Samuel for our successful projects on shared-memory atomic operations and the motivation of an endurance athlete; Matheus for our shared interest and projects around on-chip interconnects; Thomas for continuing the chip frenzy and having a sixth sense for useful technology that we had not yet been using; Lukas for many interesting discussions and critical yet constructive questions; and Maxim for taming (on-chip) interference with the stoicism of a karateka.

I was lucky to have some very talented students doing their master's thesis with me, and I particularly thank Koen for his work on the IOTLB and the early HERO software stack, Wolfgang for his work on the on-chip communication platform, and Noah for his work on the latest HERO software stack.

Last but certainly not least, I owe my friends big thanks for making life beyond work and professional passion fun and fulfilled. I cannot thank my family, especially my parents, my sister, and my grandmother, enough: not only have they always been supportive and loving but they have also nurtured my curiosity and allowed me to get fascinated by computers at an early age. The family of my wife has welcomed me like a son or brother, and I highly appreciate that as truly special. Finally, I deeply cherish my wife for her love, always being there for me, and sharing my life from everyday moments to visions and dreams.

Abstract

Heterogeneous systems on chip (HeSoCs) combine general-purpose, feature-rich multi-core *host* processors with domain-specific programmable many-core accelerators (PMCAs) to unite versatility with energy efficiency and peak performance. By virtue of their heterogeneity, HeSoCs hold the promise of increasing performance and energy efficiency compared to homogeneous multiprocessors, because applications can be executed on hardware that is designed for them. However, this heterogeneity also increases system complexity substantially.

The challenges in designing efficient and effective HeSoCs are manifold: On the hardware level, accelerator architectures need to be co-optimized with a memory architecture that enables efficient communication and synchronization within the PMCA and with the host. On the other end of the spectrum, algorithms need to be tailored to a given HeSoC without burdening application programmers with architectural intricacies. Between these two poles, many differences between host and PMCAs – such as the instruction set architecture (general-purpose, operating-system-capable vs. domain-specific, bare-metal), architectural width (64 bit vs. 32 bit), memory organization and coherence (hardware-managed coherent caches vs. software-managed scratchpad memories), and addressing scheme (virtual memory vs. physical addresses) – need to be overcome. While leading companies continue to advance their products, academic research on HeSoCs lags behind. A key reason is that simulators model HeSoCs to a limited degree at best, mainly because models of key components are too inaccurate or entirely missing, and because accurate full-stack simulation is prohibitively slow. A research platform that implements a working prototype, on the other hand, would enable efficient, collaborative, and accurate research on HeSoCs that can compete with the pace of industry.

This thesis presents the first research platform for HeSoCs where all components, from accelerator cores to application programming interface, are available under permissive open-source licenses. We begin by identifying the hardware and software components that are required in HeSoCs and by designing a representative hardware and software architecture. We then design, implement, and evaluate four critical HeSoC components that have not been discussed in research at the level required for an open-source implementation: First, we present a modular, topology-agnostic, high-performance on-chip communication platform, which adheres to a state-of-the-art industry-standard protocol. We show that the platform can be used to build high-bandwidth (e.g., 2.5 GHz and 1024 bit data width) end-to-end communication fabrics with high degrees of concurrency (e.g., up to 256 independent concurrent transactions). Second, we present a modular and efficient solution for implementing atomic memory operations in highly-scalable many-core processors, which demonstrates near-optimal linear throughput scaling for various synthetic and real-world workloads and requires only 0.5 kGE per core. Third, we present a hardware-software solution for shared virtual memory that avoids the majority of translation lookaside buffer misses with prefetching, supports parallel burst transfers without additional buffers, and can be scaled with the workload and number of parallel processors. Our work improves accelerator performance for memory-intensive kernels by up to 4×. Fourth, we present a software toolchain for mixed-data-model heterogeneous compilation and OpenMP offloading. Our work enables transparent memory sharing between a 64-bit host processor and a 32-bit accelerator at overheads below 0.7 % compared to 32-bit-only execution. Finally, we combine our contributions to a research platform for state-of-the-art HeSoCs and demonstrate its performance and flexibility in multiple case studies.

Zusammenfassung

Heterogene Ein-Chip-Systeme (HeSoCs) kombinieren universell einsetzbare, funktionsreiche, mehrkernige Hauptprozessoren mit anwendungsspezifischen, programmierbaren, vielkernigen Rechenbeschleunigern (PMCAs), um Vielseitigkeit mit Energieeffizienz und Rechenleistung zu vereinen. Durch ihre Hetorogeneität versprechen HeSoCs höhere Rechenleistung und Energieeffizienz gegenüber homogenen Mehrkernprozessoren, weil Anwendungen auf Rechenwerken ausgeführt werden können, die speziell für sie entwickelt wurden. Diese Heterogeneität erhöht die Systemkomplexität jedoch beträchtlich.

Die Herausforderungen bei der Entwicklung effizienter und effektiver HeSoCs sind vielschichtig: Auf der Schicht der Rechenwerke müssen Beschleunigerarchitekturen zusammen mit der Speicherarchitektur optimiert werden, um eine effiziente Kommunikation und Synchronisation innerhalb des PMCA sowie zwischen PMCA und Hauptprozessor zu ermöglichen. Am anderen Ende des Spektrums, auf der Schicht der Anwendungen, müssen Algorithmen auf HeSoCs angepasst werden, ohne dass sich Anwendungsentwickler mit den Details der Rechnerarchitektur auskennen. Auch zwischen jener untersten und obersten Schicht müssen viele Unterschiede zwischen Hauptprozessor und PMCA überbrückt werden. Genannt seien hierbei die Befehlssatzarchitektur (ISA), die architekturielle Breite (64 bit vs. 32 bit), die Speicherorganisation und -kohärenz (hardwareverwaltete, kohärente Zwischenspeicher stehen softwareverwalteten Blockspeichern gegenüber) und das Adressiermodell (virtualisierter Speicher steht physikalischen Adressen gegenüber). Führende Unternehmen entwickeln ihre HeSoC-Produkte stetig weiter, aber die Wissenschaft hinkt auf diesem Gebiet hinterher. Ein wichtiger Grund dafür ist, dass HeSoCs nur bedingt simuliert werden können, da Modelle von Schlüsselbausteinen zu ungenau sind oder ganz fehlen und weil die präzise Simulation aller Schichten zu langsam für die Praxis ist. Im Gegensatz zu Simulationen würde eine Forschungsplattform, die einen funktionierenden Prototypen implementiert, effiziente, kollaborative und präzise Forschung an HeSoCs erlauben, die mit der Industrie mithalten kann.

Diese Dissertation stellt die erste Forschungsplattform für HeSoCs vor, die alle Komponenten, von Rechenbeschleunigerkernen bis zur Anwendungsentwicklungsschnittstelle, quelloffen und unter liberalen Lizenzen verfügbar macht. Dazu bestimmen wir zunächst die Hard- und Softwarekomponenten, die für einen HeSoC notwendig sind, und entwickeln damit eine Referenzarchitektur. Dann entwickeln, implementieren und evaluieren wir vier zentrale Komponenten eines HeSoC, die von der Wissenschaft zuvor noch nicht detailiert beschrieben wurden. Erstens stellen wir eine modulare, topologieunabhängige, hochleistungsfähige chipinterne Kommunikationsplatform vor, die ein industrieweit eingesetztes Protokoll befolgt, das dem neusten Stand der Technik entspricht. Wir zeigen, dass diese Platform geeignet ist um breitbandige (z.B. 2.5 GHz und 1024 bit Datenbreite) Kommunikationsnetze zu konstruieren, die viele parallele Ende-zu-Ende Verbindungen unterstützen (z.B. bis zu 256 unabhängige gleichzeitige Transaktionen). Zweitens stellen wir eine modulare und effiziente Lösung zur Implementierung unteilbarer Speicheroperationen in hochskalierbaren Vielkernprozessoren vor und zeigen, dass diese Lösung in verschiedenen synthetischen und realen Anwendungsszenarios nahezu optimal linear skaliert und nur 0.5 kGE pro Kern gross ist. Drittens stellen wir eine Hardware-Software-Lösung für gemeinsam genutzten virtuell adressierten Speicher vor, welche die Mehrheit der *translation lookaside buffer* (TLB) Fehlzugriffe mit Vorabzugriffen vermeidet, parallele Sammeltransfers ohne zusätzliche Zwischenspeicher unterstützt und mit der Auslastung und der Anzahl Kerne skaliert. Unsere Lösung verbessert die Ausführungsgeschwindigkeit des Rechenbeschleunigers um bis zu Faktor 4 bei speicherintensiven Anwendungen. Viertens stellen eine Softwarewerkzeugkette für die heterogene Kompilierung mit gemischten Datenmodellen und OpenMP-Auslagerung vor. Unsere Lösung ermöglicht transparent die gemeinsame Speichernutzung zwischen einem 64-bit Hauptprozessor und einem 32-bit Rechenbeschleuniger und dies bei einer zusätzlichen Laufzeit von lediglich

0.7 % verglichen mit der isolierten 32-bit Ausführung. Zum Schluss kombinieren wir die vier Komponenten zu einer Forschungsplattform für HeSoCs, die dem Stand der Technik entsprechen, und zeigen die Geschwindigkeit und Flexibilität der Plattform in mehreren Fallstudien.

Contents

Chapter 1

Introduction

Heterogeneous computers aim to combine general-purpose computing with domain-specific, efficient processing capabilities [Hor14; Zah19; DTH20]. Such computers co-integrate a versatile multi-core *host* processor with specialized programmable many-core accelerators (PMCAs). In the case of heterogeneous systems on chip (HeSoCs), the different components are part of the same integrated circuit (IC) (also called "chip"). By virtue of their heterogeneity, heterogeneous computers promise increased performance and energy efficiency compared to homogeneous multiprocessors, because applications can be executed on hardware that is designed specifically for them.

The architectural template this thesis focuses on is shown in Fig. 1.1. A general-purpose host central processing unit (CPU) is coupled to one or multiple PMCAs via an interconnect over which they share the off-chip main memory (and input/output (I/O) peripherals). The host CPU consists of one or more general-purpose application class processing cores and has a memory hierarchy of virtually-addressed caches. The host CPU is thus capable of running a full-featured operating system (OS). The PMCAs consist of many minimal, domain-specific processing elements (PEs), which can be grouped in clusters, and have a memory hierarchy of physically-addressed, software-managed scratchpad memories (SPMs). Each PMCA may additionally be attached to an off-chip SPM managed by that PMCA. A bridge to the host allows the

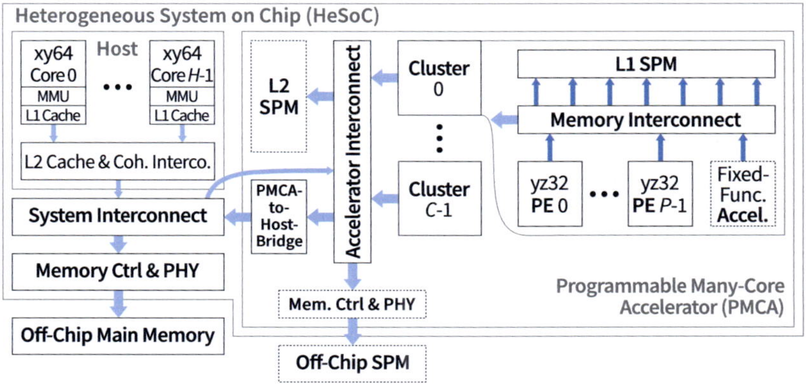

Figure 1.1: Hardware architecture template for heterogeneous systems on chip (HeSoCs) that this thesis focuses on. Components with a dashed border are optional.

PMCAs to share the main memory, and optionally some of the caches, with the host.

Programmable many-core accelerators are thus an essential component of a heterogeneous computer. PMCAs today exist in many different types and architectures, ranging from compute-optimized and general-purpose graphics processing units (GPUs) for data centers and high-performance computing (HPC) [CG20; Bly20; AMD20a] or embedded systems [AMD20b], over artificial intelligence (AI) or machine learning (ML) processors [Ouy+20; BV20; JHL20; Nor+20; Lia+19], vision processing units (VPUs) [Int20a] and digital signal processors (DSPs) [Tex19c; Cod13], to multiprocessors with a full-featured instruction set architecture (ISA) and domain-specific ISA extensions [Ros+17; ZSB20; Kal20].

Heterogeneous computing is by no means a purely hypothetical concept. On the contrary, heterogeneous computers – and HeSoCs in particular – are gaining increased commercial importance. Leading companies have introduced successful products and continue to push the technological boundaries. Products today range from internet of things (IoT) and wearable devices [Fla+18; Ter19; Ros+21] over mobile phones [Fru20b; Fru20a], tablets, and notebooks [ABW20; Fru20c] to automotive and industrial embedded systems [Ban+19; Nvi20a; Nvi20b] and warehouse-scale computing [Fun20; Hew21; Ran+21]. Redefining what computers can do with a given space, power, or energy budget, heterogeneous computing has the potential to change – hopefully positively – our lives in many ways, from work [CW15] and science [Kli20] over medical care [SBL15; Kur+16; ZHA19] and travel [Ban+19] to augmented reality [Liu+18] and leisure [Pat+15].

1.1 Challenges in Heterogeneous Computing

The challenges in designing heterogeneous computers and in making use of their conceptually immense potential are manifold.

From Hardware ...

On the hardware level, the ISA and microarchitecture of the PEs as well as the architecture of an entire accelerator have to be optimized for the target application domain. Three aspects can be highlighted:

First, the fundamental operations of an application domain define the instructions and the PE microarchitecture. For example, for many algorithms in the linear algebra and ML domain, multiply-accumulate operations (MACs) are fundamental [Sze+17]. Such fundamental operations change relatively infrequently.

Second, the data types and the memory access patterns around fundamental operations change much more rapidly than the fundamental operations themselves. With the advent of reduced- and trans-precision processing, modern accelerators in the linear algebra and ML domains have to support a wide range of precision and be able to increase performance as precision decreases – ideally for an arbitrary mix of data types that changes dynamically during execution [CV20; Mac+21].

Third, the parallel memory access patterns exhibited by an algorithm have a major influence on the memory and interconnect architecture of an accelerator. While there are fundamental patterns (such as tiling in various dimensions and stencils with various shapes and strides), those are multifaceted and change frequently as new algorithms in an application domain are developed [CSS11; Jan+11]. For these three aspects (and potentially many more), a trade-off between versatility and specialization needs to be struck in the design of every accelerator.

Next, accelerators have to be increasingly scalable, meaning the same fundamental architecture should be replicable from one instance to hundreds and the performance should ideally scale linearly with the number of PEs. While large parts of many algorithms are pleasingly parallel, any serial parts of a program limit the speed-up by parallelization [Amd67; HM08; HS11]. Atomic memory operations provide means to implement concurrent algorithms without serial parts, yet their scalable implementation in a parallel processor is not a solved problem [SBH15].

Finally, accelerators need to be co-optimized with the memory hierarchy and the interconnects of the host processor to enable efficient data sharing and communication with the host. For instance, accelerators might feature a software-managed memory hierarchy whereas host processors typically feature multiple levels of hardware-managed caches [Ban+02]. This difference results in very different on-chip communication paradigms and protocols. Address spaces and addressing modes frequently also differ between accelerators and host: The host is typically a 64-bit processor that runs an OS, which manages virtual address spaces that are private to each application [SGG18]. For accelerators, on the other hand, 32 bit are typically sufficient to address local memory, and physical addressing simplifies hardware and maximizes performance per area [Vog+15; Hao+17]. All these aspects lead to a huge design space for PMCAs, and questions on the granularity and the combination or separation of accelerators for economical reasons add another layer of complexity on top [LL03].

... to Algorithms and Applications ...

On the other end of the spectrum, algorithms and applications have to make optimal use of the available diverse hardware. This involves answering questions about task partitioning (what constitutes a unit of work), task mapping (what to execute where), task scheduling (what to execute when), and communication and synchronization (how to tile data and process it in parallel) [Sin+13; Tre18; Hua+21]. Those questions typically form complex optimization problems – some of which have already been shown to be NP-complete [Bea+02; Bea+19] – and every combination of algorithm and heterogeneous hardware potentially has a different optimal solution [Yan+10; Mor+14; LW17].

... and Many Layers Between

Between the two poles of hardware and algorithms, programming models, frameworks, runtime environments, and compiler toolchains strive to achieve high programmability and performance portability. On the lowest level, each programmable accelerator needs a compiler that generates optimized machine code from a low-level standardized language such as C or Fortran. Above that, standardized parallel programming models and runtime environments – such as OpenMP [Omp3] and OpenCL [SGS10] – allow the use of parallel compute resources without being deeply familiar with the underlying hardware. Such programming models are increasingly also developed for heterogeneous computing, where a part of a program can be specified to execute on an accelerator without having to define how exactly this *offload* from host to accelerator happens [Omp5.1; Khr21b; OpenACC3.1]. However, even such heterogeneous and parallel programming models cannot completely eliminate the need to modify and specialize application code for each target architecture. Higher-level frameworks, such as data-centric programming [Ben+19], aim to solve this problem by decoupling the description of an algorithm from its specialization to perform well on a target architecture. This decoupling can also help the aforementioned optimizers to automatically map algorithms to heterogeneous hardware.

Closing the loop to hardware, hardware-software co-design aims to specify, synthesize, optimize, and verify hardware and software together [De +02]. The concept has existed for a long time before the advent of heterogeneous computing and to date has not seen widespread adoption, presumably due to requiring deep changes to workflows, teams, and tools throughout the industry. However, in face of the challenges of heterogeneous computing, hardware-software co-design is currently reappearing in design methodologies [Tin+20] and compilers [Lat+21a; Lat+21b].

Of these manifold challenges, many, if not all, will have to be solved to bring the full potential of heterogeneous computing to bear. This requires revisiting the entire computing stack [Zah17].

1.2 The Need for an Open-Source Research Platform

A full-stack heterogeneous research platform makes revisiting the entire computing stack possible in the first place: Only a platform that integrates all essential components of a heterogeneous computer allows reproducible and falsifiable research on all components and their interfaces and interactions. This contrasts with the traditional 'two-pronged' approach, where hardware components are developed and evaluated in isolation [Che+15; Joh+11] and their impact on system-level performance is estimated through models and simulators [Bor+13; Li+09]. Compared to a heterogeneous research platform, the two-pronged approach has three significant drawbacks: First, interactions between host, accelerators, the memory hierarchy, and I/O devices are complex to model accurately, making simulations orders of magnitude slower than running prototypes. Second, even full-system simulators such as gem5 [Bin+11] model HeSoCs to a limited degree only [But+16]. For example, models of system-level interconnects or memory management units (MMUs), which dynamically influence the path from accelerators to different levels of the memory hierarchy, are missing. Third, simulations are based on assumptions. Contrary to results obtained from a working prototype, simulated results require authors to justify and reviewers to validate the underlying assumptions. Simulators that are not precisely calibrated and accuracy-validated against the simulated system are generally too inaccurate to provide significant results, and full-system simulators are particularly unreliable [AS19]. A research platform that serves as a working prototype, on the other hand, enables reproducible, falsifiable, efficient, and collaborative research and development. To perform system-level research using standard benchmarks and real-world applications, the platform must include a software stack that includes an application programming interface and a complete compiler toolchain.

Existing research platforms do not meet these requirements in their entirety. Many provide a custom accelerator on programmable logic [Gra16; KHG20], and some even couple the accelerator to a host processor that runs an OS [Man+20; Bal+20a], but none come with all required hardware and software components. We defer a full discussion of related work to § 7.3, when we have a full understanding of the challenges and components of a heterogeneous research platform.

1.3 Challenges of an Open-Source Research Platform

The main challenge in creating an open-source research platform for heterogeneous computing can be stated simply as: for many of the problems in designing and using heterogeneous computers (described in § 1.1), a solution has to be found, and that solution has to be incorporated into an open-source implementation. Those solutions do not have to be optimal. In fact, solutions that are not highly specialized and rigid but rather come as a framework that can be flexibly parametrized, modified, and extended are preferable. For this reason, it also makes sense to not innovate and optimize on every frontier but rather use existing standards and specifications wherever possible. Adhering to established standards increases compatibility and makes the platform relevant for more users. This is easier in fields with a vibrant open-source community, such as the RISC-V ISA, than in fields dominated by commercial oligopolies, such as intellectual property modules (IPs) for on-chip communication.

A second major challenge is verification. A research platform certainly does not have to be a product of commercial maturity, but it should allow its users to focus on answering research questions without being inhibited by faults in the platform. The complexity of this challenge stems from the deep interdependencies between all components. Consider a heterogeneous research platform that supports two different host processors and three different PMCA architectures. That platform will have five different compiler backends[1], six different heterogeneous toolchains[2], and six different runtime software stacks[2] (besides many other components). A change in any accelerator architecture can require re-validation of its compiler backend, two heterogeneous toolchains, and two runtime software stacks – ideally with all applications that the platform supports. The fact that many hardware components can be

[1]One for each host and one for each PMCA.

[2]One for each combination of host and PMCA.

parametrized and the value of many parameters has to be propagated throughout the software stack adds an additional layer of complexity. Adhering to standards clearly helps tackling this challenge, too, but standards at the implementation level are often prohibitively restrictive when exploring novel concepts in a research setting. However, our contributions to the verification challenge are mainly of an engineering nature, and although many thoughts and much effort went into verification, this thesis focuses on scientific contributions towards the design and use challenges.

1.4 Contributions and Publications

The focus of this thesis is to create an open-source research platform for HeSoCs. This includes the design and implementation of its main components, where they were not available in literature or as open-source implementation. Thereby, this thesis addresses the main challenge described in § 1.3 and many of the challenges described in § 1.1.

The key contributions of this thesis can be summarized as follows:

1. We present a modular, topology-agnostic, performance-scalable on-chip communication platform for a state-of-the-art, industry-standard protocol. The platform includes all components required for end-to-end on-chip communication; from DMA engines over network switches and converters to memory controllers and a last-level cache). This is the first[3] such platform described in literature, and it comes with an open-source implementation. We show that our platform can be used to build high-bandwidth (e.g., 2.5 GHz and 1024 bit full-duplex data width) on-chip communication fabrics or can scale to 1024 cores on a die, providing 32 TB/s cross-sectional bandwidth at only 24 ns round-trip latency between any two cores.

2. We present a modular, scalable, and flexible solution to implement atomic memory operations in multiprocessors with software-managed memory hierarchies. This is the first[3] such solution

[3] All 'first' and 'novel' statements in this section refer to the time of publication of the corresponding article and are to the best of our knowledge.

described in literature, and it comes with an open-source implementation. We show that our solution scales linearly in area with the number of cores (only 0.5 kGE per core, measured in 22 nm FDSOI) and allows the throughput of important concurrent algorithms to scale linearly with the number of cores until the memory bandwidth is saturated.

3. We present an efficient solution for sharing virtual memory between host and PMCAs in HeSoCs. Our solution includes three novel[3] concepts. Applied together, those concepts allow our solution to improve PMCA performance for memory-intensive kernels by up to 4× and by up to 60 % for irregular and regular memory access patterns, respectively, compared to the previous state of the art [Vog+17].

4. We present the first[3] mixed-data-model compiler, supporting arbitrary address widths on host and PMCA. To hide the inherent complexity and enable high programmer productivity, the compiler supports transparent offloading on top of OpenMP. Our evaluation on a 64+32-bit HeSoC shows that memory can be shared between host and PMCA without programmer intervention at overheads below 0.7 % compared to 32-bit-only execution, enabling mixed-data-model heterogeneous computers to execute at near-native performance.

5. We present the first[3] full-stack research platform for HeSoCs, including the aforementioned contributions as components. We study four current research topics in heterogeneous computing and provide quantitative insights on the level of applications, toolchains, system architecture, and accelerator architecture, and we make the platform available to the community under permissive open-source licenses.

The content of this thesis and the main contributions have been published as follows (all peer-reviewed):

[Kur+17] **A. Kurth**, P. Vogel, A. Capotondi, A. Marongiu, and L. Benini, "HERO: Heterogeneous embedded research platform for exploring RISC-V manycore accelerators on FPGA", in *Proceedings of the*

1st Workshop on Computer Architecture Research with RISC-V, CARRV '17, Boston, MA, USA, October 2017.

[Kur+18a] **A. Kurth**, P. Vogel, A. Marongiu, and L. Benini, "Scalable and efficient virtual memory sharing in heterogeneous SoCs with TLB prefetching and MMU-aware DMA engine", in *Proceedings of the 36th IEEE International Conference on Computer Design*, ICCD '18, Orlando, FL, USA, October 2018.

[Kur+18b] **A. Kurth**, A. Capotondi, P. Vogel, L. Benini, and A. Marongiu, "HERO: An open-source research platform for HW/SW exploration of heterogeneous manycore systems", in *Proceedings of the 2nd Workshop on AutotuniNg and aDaptivity AppRoaches for Energy Efficient HPC Systems*, ANDARE '18, Limassol, Cyprus, November 2018.

[Kur+20a] **A. Kurth**, S. Riedel, F. Zaruba, T. Hoefler, and L. Benini, "ATUNs: Modular and scalable support for atomic operations in a shared memory multiprocessor", in *Proceedings of the 57th ACM/IEEE Design Automation Conference*, DAC '20, June 2020. Best Paper Nominee (6/228). HiPEAC Paper Award.

[Kur+20b] **A. Kurth**, K. Wolters, B. Forsberg, A. Capotondi, A. Marongiu, T. Grosser, and L. Benini, "Mixed-data-model heterogeneous compilation and OpenMP offloading", in *Proceedings of the 29th ACM International Conference on Compiler Construction*, CC '20, San Diego, CA, USA, February 2020.

[Kur+22] **A. Kurth**, W. Rönninger, T. Benz, M. Cavalcante, F. Schuiki, F. Zaruba, and L. Benini, "An open-source platform for high-performance non-coherent on-chip communication", in *IEEE Transactions on Computers*, vol. 71, no. 8, pp. 1794-1809, August 2022.

[KFB22] **A. Kurth**, B. Forsberg, and L. Benini, "HEROv2: Full-stack open-source research platform for heterogeneous computing", in *IEEE Transactions on Parallel and Distributed Systems*, July 2022 (early access).

The author has further contributed to related research projects, which are not directly covered in this thesis, while working towards his degree. The results have been published as follows (all peer-reviewed):

[Vog+17] P. Vogel, **A. Kurth**, J. Weinbuch, A. Marongiu, and L. Benini, "Efficient virtual memory sharing via on-accelerator page table walking in heterogeneous embedded SoCs", in *ACM Transactions on Embedded Computing Systems*, September 2017.

[Di +19] S. Di Girolamo, K. Taranov, **A. Kurth**, M. Schaffner, T. Schneider, J. Beránek, M. Besta, L. Benini, D. Roweth, and T. Hoefler, "Network-accelerated non-contiguous memory transfers", in *Proceedings of the 2019 ACM International Conference for High Performance Computing, Network, Storage and Analysis*, SC '19, Denver, CO, USA, November 2019.

[Cav+20] M. Cavalcante, **A. Kurth**, F. Schuiki, and L. Benini, "Design of an open-source bridge between non-coherent burst-based and coherent cache-line-based memory systems", in *Proceedings of the 17th ACM International Conference on Computing Frontiers*, CF '20, May 2020.

[For+20] B. Forsberg, M. Mattheeuws, **A. Kurth**, A. Marongiu, and L. Benini, "A synergistic approach to predictable compilation and scheduling on commodity multi-cores", in *Proceedings of the 21st ACM SIGPLAN/SIGBED Conference on Languages, Compilers, and Tools for Embedded Systems*, LCTES '20, June 2020.

[Sch+20] F. Schuiki, **A. Kurth**, T. Grosser, and L. Benini, "LLHD: A multi-level intermediate representation for hardware description languages", in *Proceedings of the 41st ACM SIGPLAN Conference on Programming Language Design and Implementation*, PLDI '20, June 2020. HiPEAC Paper Award.

[Mat+21] M. Mattheeuws, B. Forsberg, **A. Kurth**, and L. Benini, "Analyzing memory interference of FPGA accelerators on multicore hosts in heterogeneous reconfigurable SoCs", in *Proceedings of the 2021 IEEE/ACM Design, Automation and Test in Europe Conference*, DATE '21, March 2021.

[Di +21] S. Di Girolamo, **A. Kurth**, A. Calotoiu, T. Benz, T. Schneider, J. Beránek, L. Benini, and T. Hoefler, "A RISC-V in-network accelerator for flexible high-performance low-power packet processing", in *Proceedings of the 48th IEEE/ACM Annual International Symposium on Computer Architecture*, ISCA '21, June 2021.

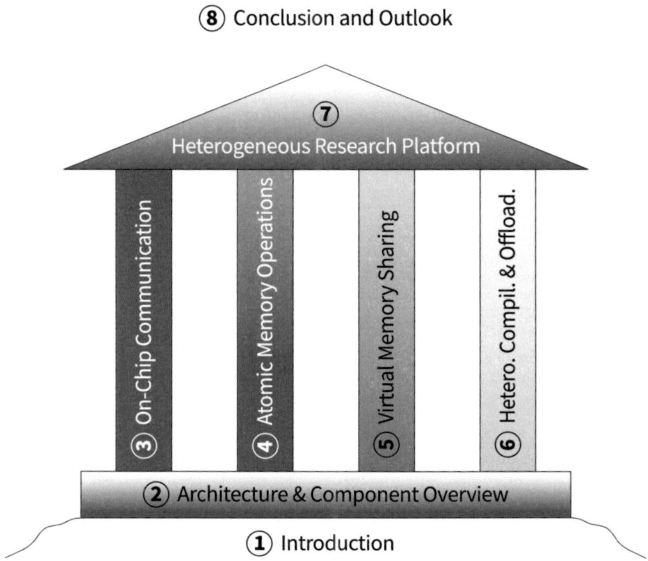

Figure 1.2: Structure of this thesis. Each numbered element stands for one chapter. The darker the fill of an element, the more it relates to hardware, and the lighter the more to software. Chapters 3 to 6 describe components that are mostly independent of each another.

1.5 Outline

The remainder of this thesis is organized as shown in Fig. 1.2. Chapter 2 introduces the architecture of our heterogeneous research platform, gives an overview of its components, presents a first prototype implementation, and discusses its limitations. Chapters 3 to 6 then resolve those limitations through contributions to four important components of heterogeneous computing. Those chapters are sorted by their focus on hardware or software; Chapter 3 focuses exclusively on hardware and Chapter 6 focuses almost exclusively on software. Chapter 3 presents a modular, topology-agnostic, high-performance on-chip communication platform with components to build and link subnetworks with customizable bandwidth and concurrency properties. Chapter 4 presents a modular solution to implement atomic memory operations, called

Atomic Unit (ATUN), which can be placed flexibly at different levels in the memory hierarchy. Chapter 5 presents a solution for virtual memory sharing that avoids the majority of translation lookaside buffer (TLB) misses with prefetching, supports parallel direct memory access (DMA) transfers without additional buffers, and can be scaled with the workload and number of parallel processors. Chapter 6 presents a mixed-data-model heterogeneous compiler including transparent offloading with OpenMP. The components described in Chapters 3 to 6 are mostly independent of each another, so those chapters could be read in any order. Chapter 7 presents a full-stack open-source research platform for state-of-the-art heterogeneous computing. This platform includes all components presented in previous chapters and discusses and evaluates their system-level integration. The chapter studies four current research topics in heterogeneous computing and provide quantitative insights on the level of applications, toolchains, system architecture, and accelerator architecture. The chapter also proposes and evaluates a novel solution to one of the most pressing problems in heterogeneous computing: how to relieve the programmer of the burden of specializing an algorithm to the memory hierarchy of an accelerator. Chapter 8 concludes the thesis and gives an outlook on the opportunities and challenges ahead.

Chapter 2

Architecture and Components of a Heterogeneous Research Platform

This chapter introduces the hardware and software architecture of our heterogeneous research platform, gives an overview of its components, presents a first prototype implementation, and discusses its limitations.

2.1 Hardware Architecture

The hardware architecture of a heterogeneous research platform must be as flexible – that is, not only parametrizable but also modifiable and extensible – as possible. However, the hardware should also adhere to existing industry standards whenever possible, to maximize the surface of compatibility and make the platform relevant for many users. The *Zynq* family of systems on chip (SoCs) by Xilinx [Xil16] combines a hard-macro ARM Cortex-A host CPU with a field-programmable gate array (FPGA) (also called "programmable logic (PL)") on a single die. This combination of an industry-standard host and its memory hierarchy and memory controllers with an FPGA, on which custom PMCAs can be implemented, offers an ideal starting ground for a research platform on HeSoCs. The idea of combining an industry-standard host implemented as hard macro with custom PMCAs implemented on an FPGA is not limited to Zynq-like SoCs, however. Product development platforms such as the Juno ARM Development Platform (Juno ADP) implement a Cortex-A host CPU and a low-latency chip-to-chip interface, which can be connected to an FPGA on a separate IC. This separation over two ICs allows more silicon area for the hard-macro host and the soft-macro PMCA.

Using FPGAs to implement the PMCAs has more advantages than just reprogrammability. First, many come with physical layer (PHY) and media access control layer IPs that are not available for research and development ICs that will not directly results in a commercial tape-out. Examples include off-chip memories such as double data rate (DDR) dynamic random-access memory (DRAM) and High Bandwidth Memory (HBM) or Ethernet. Second, vendor-provided debug and trace solutions make it relatively simple to inspect signals on an FPGA just as one would in register-transfer level (RTL) simulation. Third, FPGA vendors make many on-chip IPs available at no extra charge with their devices. Of particular interest for implementing HeSoCs are IPs for standard on-chip communication protocols such as Advanced eXtensible Interface (AXI). Those vendor-provided IPs are not open-source, but the liberal license of the AXI specification allows open-source implementations. Thus, IPs provided by the FPGA vendor can be used for a first prototype and later replaced with an open-source implementation that allows full design transparency and modifiability.

2.1.1 PMCA ISA and Architecture

An important design aspect of the PMCA is its ISA, because it is the interface between software and hardware and ultimately determines their usability and performance in the system. The RISC-V ISA [Wat16] has built considerable momentum in the community [Cel+15; Gau+17; Zim+16] because it is an open standard and designed in a modular way: a small set of base instructions is accompanied by standard extensions and can be further extended through custom instructions [Kan16]. This allows computer architects to implement the extensions suitable for their target application. Moreover, the ISA is suitable for various types of processors from tiny microcontrollers [Tra+16] to high-performance super-scalar out-of-order cores [Cel18], because it does not specify implementation properties. Combined, these characteristics make RISC-V an interesting candidate for specialized PMCAs.

There are many different PMCA architectures, such as Kalray MPPA [Din+13], KiloCore [Boh+17], STHORM [Mel+12], Epiphany [Olo16], and PULP [Ros+14b]. PULP (short for *Parallel Ultra Low Power* platform) is an architectural template for scalable, energy-efficient processing that combines an explicitly-managed memory hierarchy, ISA extensions and compiler support for specialized DSP instructions, and energy-efficient cores operating in parallel to meet processing performance requirements. PULP is a silicon-proven [Gau+17], open [Tra+16] architecture implementing the RISC-V ISA, and it can cover a wide range of performance requirements by scaling the number of cores or adding domain-specific extensions. Thus, it is well suited to serve as a baseline PMCA in research on HeSoCs.

2.1.2 Heterogeneous Architecture

The hardware architecture of our first-generation heterogeneous research platform (HEROv1), which is shown in Fig. 2.1, is a result of the aforementioned design choices (a hard-macro host combined with soft-macro PMCAs on an FPGA, RISC-V as PMCA ISA, and PULP clusters as PMCA template). The host consists of ARM Cortex-A cores attached to a coherent interconnect. Each core includes private hardware-managed caches and an MMU. The host shares main memory with the PMCAs through the system interconnect, which is coherent to the caches of the host. To overcome scalability limitations, the PMCA features a cluster-based design and relies on multi-banked, software-managed SPM and DMA engines instead of data caches. The 32-bit RISC-V PEs within a cluster primarily operate on data present in the cluster-local L1 SPM to which they connect through a low-latency logarithmic interconnect. The PEs use the cluster-internal DMA engine to copy data between the local L1 SPM and remote SPMs or shared main memory. Transactions to main memory pass through a hybrid input/output memory management unit (IOMMU), implemented by

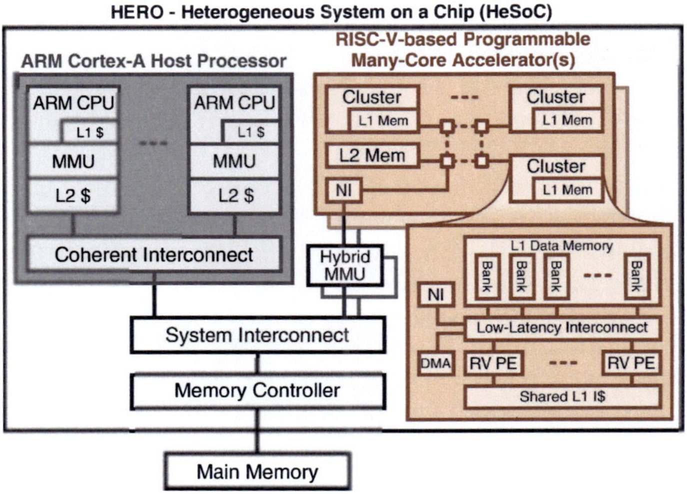

Figure 2.1: Hardware architecture of HEROv1.

the Remapping Address Block (RAB) [Vog+15], which performs virtual-to-physical address translation based on the entries of an internal table, similar to the MMUs of the host CPU cores. This lightweight hardware block is manged in software directly on the PMCA [Vog+17]. The host and the PMCA can thus efficiently share virtual address pointers. This enables shared virtual memory (SVM), which substantially eases overall system programmability and enables efficient sharing of linked data structures in the first place. The hybrid IOMMU is connected as an I/O-coherent device on the system interconnect, which allows the PMCA to access the shared main memory, optionally also coherently with the caches of the host.

2.1.3 PMCA Configurability

The PMCA is highly configurable. Table 2.1 gives an overview of the most important among the supported synthesis parameters. Besides the number of clusters and the number of PEs and SPM banks per cluster, the 32-bit RISC-V PEs themselves can be configured to trade off hardware resources and computing performance. The single-precision

Component	Options
# Clusters	1, 2, 4, **8**
# PEs per cluster	2, 4, **8**
floating-point unit (FPU)	Private, shared, **off**
Integer DSP unit, divider, multiplier	**Private**, shared
L1 SPMs # banks	4, 8, **16**
L1 SPMs size [KiB]	32, 64, 128, **256**
L2 SPM size [KiB]	32, 64, 128, **256**
Instruction cache design	**Single-** or multi-ported
Instruction cache size [KiB]	2, 4, **8**
Instruction cache # banks	2, 4, **8**
RAB L1 TLB size	4, 8, 16, **32**, 64
RAB L2 TLB size	0, 256, 512, **1024**, 2048
RAB L2 TLB associativity	16, **32**, 64
RAB L2 TLB # banks	1, 2, **4**, 8

Bold and underlined values refer to implementations discussed in § 2.3.1.

Table 2.1: Configuration options for HEROv1's PULP PMCA.

FPU can be private, shared among multiple PEs within a cluster, or completely disabled. Similarly, the integer DSP extension unit, the divider, and the multiplier can be private or shared. In addition, different designs for the shared instruction cache (e.g., single- or multi-ported) can be selected. The RAB is also configurable: the number of TLB entries and levels as well as the architecture of the second-level TLB can be adjusted.

2.2 Software Architecture

The software architecture of a heterogeneous research platform must be designed for first-class integration of PMCAs. Ideally, application software would be portable over different HeSoC designs and optimized by the toolchain to exploit the hardware architecture of a specific design. The software architecture also plays a crucial role to exploit the computational synergy between PMCAs and host: it must be designed to enable efficent data sharing between components with vastly different memory architectures, as caches and virtual memory in the host contrast with physically-addressed, software-managed SPMs in PMCAs.

As for the programming framework, OpenMP [DM98] has been established as the de-facto standard programming framework for *homogeneous* shared memory parallel programming. It is very effective at expressing single program multiple data (SPMD) loop-level parallelism through compiler directives, but it was originally not designed for *heterogeneous* computing. Extensions to the OpenMP standard were proposed by academia [Dur+11; Mar+15] and by the industry [Mit+14]. The synthesis of these works has been incarnated into OpenMP 4.0 [Omp4.0]. Since that version, OpenMP allows to offload work from a host processor to a PMCA through the `target` directive, which is already being used to program GPUs in HPC [MMG16]. This makes OpenMP a suitable candidate for a unified programming interface of HeSoCs.

Efficient data sharing between host and PMCAs is crucial in HeSoCs. However, OpenMP just knows copy-based offloading, where the host copies data from virtually-addressed, cached host memory to physically-adressed SPMs in a PMCA. This puts a daunting task on application developers: they need to deal with cache flushes, virtual-to-physical address translation, and DMA transfers of properly-sized data tiles. Beyond violating programmability requirements, those operations are very costly, make PMCAs highly dependent on the host, and often kill performance [Cho+16]. In contrast, SVM enables to share data by simply passing a pointer from host to a PMCA. This works when PMCAs can access the shared main memory coherently with the caches of the host (e.g., through an Accelerator Coherency Port (ACP) [Xil16] or as I/O-coherent slave node (SN-I) in Coherent Hub Interface (CHI) [CHI-D]) and have a IOMMU to translate virtual addresses at run time. IOMMUs that support large numbers of parallel accesses common for PMCAs were long available only at prohibitive hardware costs, but recently developed *hybrid IOMMUs* [Vog+17] have changed this.

The software stack of HEROv1, which is shown in Fig. 2.2, solves the described system-level challenges of programmability and data sharing, and it enables fast and easy-to-use heterogeneous programming. Its components seamlessly integrate the PMCA into the host system and allow for transparent accelerator programming using the OpenMP programming interface and SVM hardware capability. In single-source heterogeneous programming, execution of an application starts on

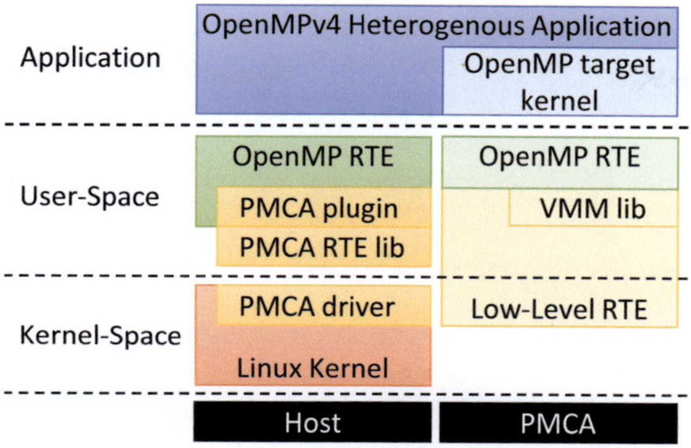

Figure 2.2: Software architecture of HEROv1.

the host, and computations can be offloaded to a PMCA simply by encapsulating a code segment into an OpenMP `target` region. The actual offload from the host is then taken care of by the OpenMP runtime environment (RTE) and afterwards by the lower levels of the software stack that is deployed on the PMCA.

2.2.1 Heterogeneous Cross Compilation

To allow the host OpenMP RTE to perform an offload to the PMCA, the `target` code region must be outlined by the host compiler, compiled by the PMCA-specific target compiler, linked against PMCA-specific libraries, and embedded into the final host *fat binary*. HEROv1's toolchain is based on the GNU GCC 7 compiler, which already supports the outlining of OpenMP `target` regions for Heterogeneous System Architecture (HSA), Nvidia PTX, and Intel Xeon Phi devices. At the end of interprocedural analysis (IPA), the GCC intermediate representation (IR) of all offloaded functions is streamed out into a link-time optimization (LTO) object. When the linker is executed, its LTO Wrapper enables link-time recompilation if at least one object file contains LTO sections. HEROv1's GNU GCC extends the offloading capability of the compiler to RISC-V-based accelerators. To enable such new devices, we extended the LTO Wrapper to execute a new `mkoffload` for HEROv1 that (i) invokes the specific ISA backend for the PMCA (in this case RISC-V), (ii) statically links the target-specific libraries (including the OpenMP runtime library), (iii) fills the Offload Table, which stores all target hooks for the outlined functions in the host binary, and (iv) packs everything into an Offload Image that is loaded to the PMCA at run time.

2.2.2 Heterogeneous OpenMP Runtime

The GNU OpenMP library, `libgomp`, supports plugins to bind target-specific implementations of runtime functions to its generic application programming interface (API). HEROv1's customized GNU GCC includes two `libgomp` plugins for the PMCA to implement two different offloading schemes: copy-based and zero-copy with SVM. In copy-based offloads, all shared data is copied to a contiguous, unpaged, uncached memory region and the PMCA is given the physical address into that region. In SVM-based offloads, the PMCA accesses host virtual addresses through a hybrid IOMMU and instrumented load/store operations (see § 2.2.3).

The first time a `target` code region is going to be executed, the host OpenMP runtime loads the dynamic shared object (DSO) containing the binary of all offloaded functions together with the accelerator-side OpenMP runtime onto the target PMCA. For every `target` code region with copy-based semantics, the host OpenMP runtime copies the shared data to contiguous memory, passes a physical address to the PMCA, and copies data back after the PMCA has finished executing. For every `target` code region with SVM-based offload semantics, the host OpenMP runtime simply passes virtual addresses to the shared data to the PMCA. All these host-to-PMCA interactions go through the host runtime library (see § 2.2.4).

2.2.3 Compiler Support for Hybrid-MMU-Based SVM

With a hybrid IOMMU, loads and stores to SVM by PEs in the PMCA can fail if the accessed virtual address misses in the TLB. As this semantic deviates from standard load and store, the compiler instruments [Vog+15] accesses to SVM with an additional read of a register, through which a PE is informed whether its last SVM access was successful. The compiler transforms memory accesses inside a `target` code region based on the data sharing context: During the OpenMP expansion pass, the compiler annotates all shared variables as candidates for instrumentation. In a static single assignment (SSA) pass, it traverses *use-def* chains to determine which uses of the annotated variables need to be instrumented. Scalar variables can then directly be instrumented, while pointer variables require an additional *escape analysis* to determine when and how a pointer dereference is propagated to instrument accesses through the propagated value.

2.2.4 Host Runtime Library and Linux Driver

The host RTE library interfaces the host-side OpenMP runtime with the Linux driver. In addition, it reserves all virtual addresses overlapping with the physical address map of the PMCA. This is required as any access of the PMCA to a shared variable located at such an address would not be routed to SVM but instead to its internal SPMs or memory-mapped registers. The driver handles low-level tasks such as interrupt handling, synchronization between PMCA and host, host cache maintenance, operation of the system-level DMA engine (e.g., to offload the PMCA binary), and initially setting up the hybrid IOMMU to give the PMCA access to the page table of the heterogeneous user-space application. The PMCA is accessed by the RTE library as a memory-mapped device, which allows for low-latency host-to-PMCA communication.

2.2.5 PMCA Virtual Memory Management Library

Having access to the page table of the heterogeneous user-space application, the PMCA can operate its virtual memory hardware autonomously. A virtual memory management (VMM) library [Vog+17] on the PMCA abstracts away differences between host architectures and IOMMU configurations and provides a uniform API to explicitly map pages and handle TLB misses. When a core accesses virtual memory through the hybrid IOMMU, the corresponding address translation may be missing in the TLB. In this case, the core that caused the miss goes to sleep and the miss is added to a queue in the L1 SPM. To handle a miss, the VMM library dequeues it, translates its virtual address to a physical one by walking the page table of the host user-space process, selects a TLB entry to replace and configures it accordingly, and wakes up the core that caused the miss. The VMM library is compatible with any host architecture supported by the Linux kernel.

2.3 Prototype Platform and Evaluation

In this section, we describe the two carrier platforms of HEROv1 (§ 2.3.1) and evaluate the design choices for our heterogenous research platform. In § 2.3.2 we explore the scaling of parallel execution and memory hierarchy usage and the limitation of the on-chip network. In § 2.3.3, we show the impact of SVM on the total PMCA run time. Finally, in § 2.3.4, we evaluate the speedup through offloading from a dual-core ARM host to an eight-core PMCA.

2.3.1 Carrier Platforms

HEROv1 has been implemented on two different carrier platforms.

Juno ARM Development Platform (Juno ADP) The Juno ADP features an ARMv8-based, multi-cluster host CPU (two A57 and four A53 cores), a Mali-T624 GPU, and 8 GiB of DDR3L DRAM. In addition, the SoC offers a low-latency AXI chip-to-chip interface (TLX-400) connecting to a Xilinx Virtex FPGA, through which 4 to 8 PMCA clusters on the FPGA can access the shared DRAM coherently with the caches of the host. The ARMv8 host CPU runs 64-bit Linaro Linux 4.5 with a 64-bit root filesystem (both `aarch64-linux-gnu`) generated using the OpenEmbedded build system. We have configured the root filesystem to have `multilib` support, such that the host can also execute 32-bit binaries (`arm-linux-gnueabihf`) in ARMv7 mode, which guarantees compatibility of data and pointer types between the host and the 32-bit PMCA architecture in heterogeneous applications.[1]

[1]This compatibility could also be achieved by running the application binary in ILP32 mode, which would allow the host to use ARMv8-specific CPU features. However, the support for ILP32 is still experimental in Linaro.

Xilinx Zynq ZC706 Evaluation Kit (ZC706) The Xilinx Zynq-7045 SoC found on the ZC706 combines an ARMv7, dual-core A9 host CPU with a Kintex FPGA on a single IC. The two subsystems are connected through a set of low-latency AXI interfaces and share 1 GiB of DDR3 DRAM. Using the ACP, the single PMCA cluster instantiated in the FPGA can also coherently access data from the data caches of the host. The main advantages of the ZC706 is higher availability and better affordability compared to the Juno ADP. The 32-bit ARMv7 host CPU runs Xilinx Linux 3.18 with a root filesystem generated using Buildroot.

		Juno ADP		ZC706	
All Clusters	LUT	936 k	76 %	128 k	59 %
	FF	450 k	18 %	43 k	10 %
	DSP	384	18 %	48	5 %
	BRAM	1152	89 %	384	70 %
Top Level and Host Interface	LUT	70 k	6 %	24 k	11 %
	FF	61 k	2 %	26 k	6 %
	DSP	0	0 %	0	0 %
	BRAM	75	6 %	71	13 %

Table 2.2: PMCA FPGA resource utilization

FPGA resource utilization The FPGA resource utilization of the PMCA on the two carrier platforms in terms of lookup table (LUT) slices, flip-flops (FFs), DSP slices, and block random access memory (BRAM) cells is shown in Table 2.2. The table lists both the absolute and the relative usage of the clusters and the top-level module containing also the host interfaces. The configuration parameters selected for implementation are highlighted as bold and underlined text in Table 2.1 for the Juno ADP and the ZC706, respectively. The clusters dominate resource usage: 8 clusters on the Juno ADP and 1 cluster on the ZC706 account for more than 90 % and 80 % of the total resource usage, respectively. While LUT and DSP slices scale linearly from the single cluster on the ZC706 to the 8 clusters on the Juno ADP, BRAMs and FFs behave differently due to different instruction cache designs: the larger, single-ported cache on the Juno ADP uses more FFs and less BRAMs per cluster than the multi-ported cache on the ZC706. Neither configuration includes FPUs, and the integer data path alone uses relatively little DSP slices, even though it supports multiplication and division. The top-level configuration is identical for both platforms, with the exception of the different interfaces to the host and the number of clusters, which enlarges the registered SoC bus. On both platforms, the available LUTs and BRAMs are the limiting factors. The PMCA can be clocked at 31 MHz and 57 MHz on the Juno ADP and the ZC706, respectively. The difference is due to the denser utilization of the Juno ADP and the fact that the Virtex FPGA of the Juno ADP consists of multiple dies connected through stacked silicon interconnects.

2.3.2 Case Study: Parallel Speedup Analysis

To demonstrate the parallel execution and data transfer capabilities of the PMCA, we use a matrix-matrix multiplication benchmark on the Juno ADP. The computations for calculating the product of two matrices, $C = AB$, are distributed over the clusters by tiling A and C row-wise. Each cluster iterates over its rows and parallelizes each row block-wise over its cores: it transfers a row of A and a column of B from the DRAM to its local L1 SPM banks, computes the resulting row of C into its L1 SPM, and transfers the resulting row to the DRAM.

The speedup achieved when parallelizing the workload over multiple clusters is shown in Fig. 2.3. In the baseline (leftmost bar), a single cluster performs the work. The bars to the right of the baseline are for two, four, six, and eight clusters. Parallelizing execution over two, four, and six clusters leads to ideal speedups compared to the baseline. For eight clusters, the interconnect between the clusters becomes the bottleneck to data transfers and limits the speedup to ca. 5 % below the ideal value. In the evaluated implementation, the interconnect is a bus, which provides low latency but no scalable bandwidth. A different on-chip network topology would scale better in bandwidth and could thus, depending on the target workload, reduce the overall execution time by supporting parallel data transfers for even more PEs. However, the used network IPs do not support different topologies or concurrency

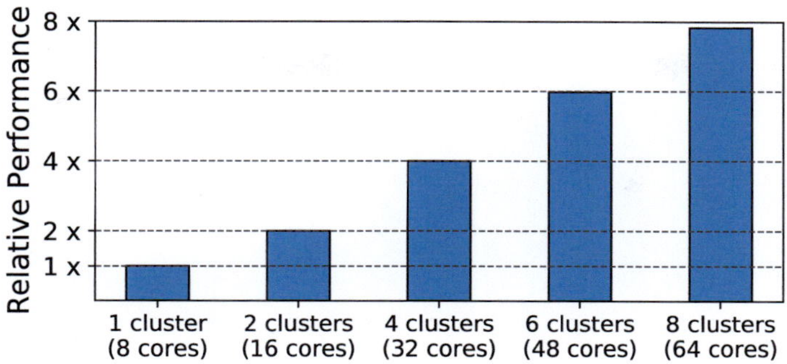

Figure 2.3: Overall execution speedup by parallelizing matrix-matrix multiplication on the PMCA implemented on the Juno ADP.

and bandwidth configurations. This also highlights the need for a more flexible, high-performance on-chip communication framework.

2.3.3 Case Study: Virtual Memory Performance Analysis

SVM support in the PMCA is essential for efficient data sharing between host and the PMCA: Without SVM, data must be copied to and from a dedicated, physically-contiguous, uncached memory section before and after accelerator execution, respectively. This copy operation depends on the data structure and may be very complex; e.g., the values of all pointers in a linked data structure must be changed. With SVM, offloading simply means passing a pointer.

The run time of different benchmarks executed on the PMCA of the Juno ADP with (orange, right bar in a pair) and without SVM (blue, left bar in a pair) is shown in Fig. 2.4. The run time is broken down into offload time, i.e., the time it takes the host to offload the computation and prepare the data for the PMCA, and the actual kernel execution time on the PMCA. All times are normalized to the total run time without SVM. **PageRank (a)** is a well-known algorithm for analyzing the connectivity of graphs and is used, e.g., for ranking web sites. It is based on a linked data structure, which makes copy-based

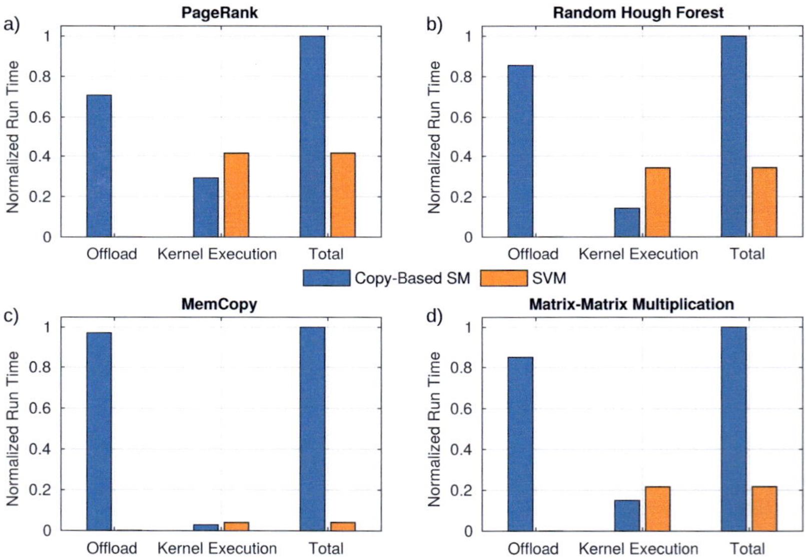

Figure 2.4: Offload and kernel execution time for different benchmarks with and without SVM support on the Juno ADP.

offloading expensive because the host must modify many pointers. With SVM, virtual addresses must be translated at run time. This causes a run time overhead if translations are not in the TLB of the RAB. Nonetheless, the offload time of copy-based SM dominates, and SVM reduces the run time by nearly 60 %. **Random Hough Forests (b)** consist of multiple binary decision trees and are used, e.g., for image classification. The trees have a very large memory footprint, but only a part of them is accessed, depending on the input data. With SVM, the PMCA can readily access the entire trees by performing the necessary address translations at run time. With copy-based SM, the trees must be made available to the PMCA in their entirety before classification can start. This leads to a lot of data being copied by the host that is never accessed by the PMCA. SVM reduces the run time by more than 60 %. **MemCopy (c)** simply copies a large array into the PMCA and back to memory. This benchmark is representative of streaming applications that require the PMCA to perform only little work. With copy-based SM, letting the host copy data to and from the physically contiguous, uncached memory to prepare the offload clearly dominates the run time. In contrast, the PMCA benefits from high-bandwidth DMA transfers. SVM removes the need for data copying by the host, reducing the total run time by more than 95 %. The **matrix-matrix multiplication benchmark (d)** involves three matrices stored in arrays, thus shows the same basic behavior as MemCopy. However, as the PMCA performs computations while traversing the data, the copy-based offload becomes a lesser part of the total run time. In this case, SVM reduces the total run time by nearly 80 %.

2.3.4 Case Study: Heterogeneous Speedup Analysis

We evaluated the speedup through offloading from the dual-core ARM host CPU to the eight-core RISC-V PMCA on the ZC706 implementation with two benchmarks from different application domains that frequently demand acceleration: matrix-matrix multiplication for signal processing and the Advanced Encryption Standard (AES) block cipher for cryptography. Both benchmarks are written in a single source file, in which we annotate the benchmark kernel with different OpenMP directives to show the performance difference of each variant. We measure the run time of each execution variant on the host using clock_gettime(CLOCK_MONOTONIC_RAW). The measurements contain *all* parts of an offload, including synchronization and transfer of data and parameters between host and PMCA. Static data on host and PMCA are initialized at the start of the benchmark application, before the first measurement.

Matrix-Matrix Multiplication In this benchmark, two square matrices a and b are multiplied into square matrix c. Each matrix has 128 by 128 32-bit elements and thus takes 64 KiB of memory. The C code for the application kernel is

```
for (unsigned i = 0; i < 128; i++) {
  for (unsigned j = 0; j < 128; j++) {
    unsigned sum = 0;
    for (unsigned k = 0; k < 128; k++) {
      sum += a[128 * i + k] * b[128 * k + j];
    }
    c[128 * i + j] = sum;
  }
}
```

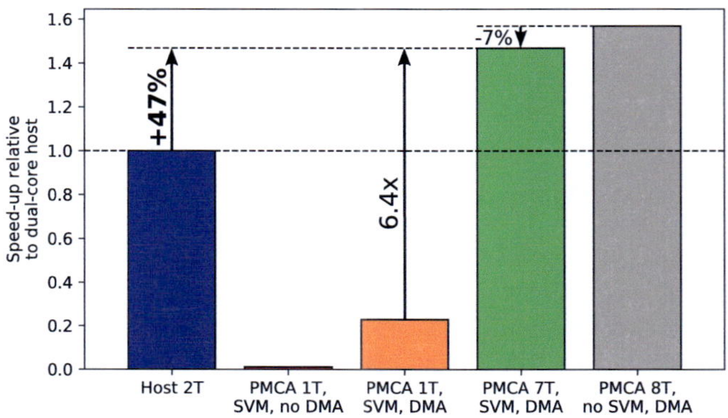

Figure 2.5: Performance of different matrix-matrix multiplication implementations on the ZC706, relative to a baseline using the dual-core ARM host CPU.

Fig. 2.5 shows the performance of different implementations, relative to the baseline (leftmost, blue bar) in which the dual-core ARM CPU runs the kernel code annotated with

```
#pragma omp parallel for firstprivate(a, b, c) collapse(2)
```

That is, the multiplication is parallelized over the rows of `a`. Offloading the kernel to the PMCA is as simple as annotating the kernel code with

```
#pragma omp target map(to: a, b) map(from: c)
```

The PMCA then directly accesses each word of any matrix through pointers to SVM. However, as the PMCA does not feature caches, performance drops drastically (second, red bar). Since the PMCA is designed to operate on its local SPM, the first optimization is to allocate buffers in SPM before the loop and to insert DMA transfers into the loop. With this, the performance improves to what a single PE can handle (third, orange bar). The kernel is now compute-bound and can be parallelized with the same annotation used to parallelize the ARM code. With this, 7 PEs process the multiplication in parallel – when using SVM, one PE is statically allocated to manage the hybrid MMU in this implementation – leading to a net performance improvement of 6.4×. Compared to a bare-metal, hand-tuned C implementation, where 8 PEs operate on buffers in SPM and DMA transfers run between SPMs and physically-addressed shared memory (rightmost, gray bar), the run-time overhead is only 7 %. Compared to the dual-core ARM implementation, on the other hand, the speed-up is 47 %, even though the PMCA on the FPGA runs at only 50 MHz. For a HeSoC IC, frequencies of 2 GHz and 800 MHz for host and PMCA, respectively, would be more realistic. In that case, this kernel would not saturate the memory bandwidth (even of the Z-7045, which is around 300 MB/s), and offloading it to the PMCA could bring a speed-up of ca 9×.

AES Block Cipher We use a popular, small C implementation [kok18] of the AES block cipher to encrypt 1024 different ciphertexts, each 128 B long, with 128-bit keys in CBC mode. The C code for the application kernel is

```
for (unsigned i = 0; i < 1024; i++) {
  AES_CBC_encrypt_buffer(ctx[i], buf[i], 128);
}
```

where `ctx` are AES contexts containing initialization vectors and keys and `buf` are the buffers that contain the plaintext before encryption and the ciphertext after encryption.

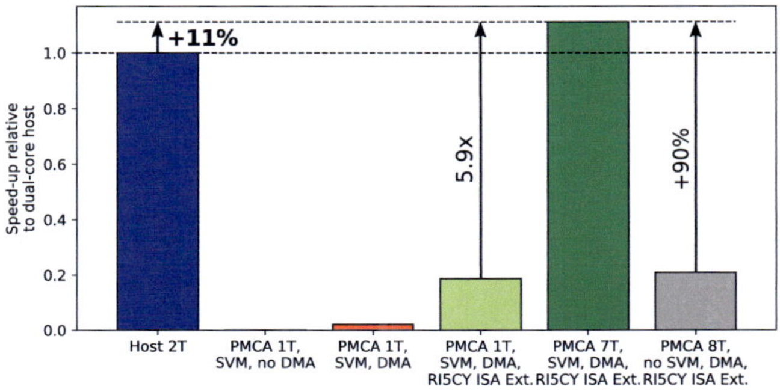

Figure 2.6: Performance of different AES implementations on the ZC706, relative to a baseline using the dual-core ARM host CPU.

Fig. 2.6 shows the performance of different implementations, relative to the baseline (leftmost, blue bar) in which the dual-core ARM CPU runs the kernel code annotated with

```
#pragma omp parallel for firstprivate(ctx, buf)
```

The second, dark red bar shows the offload to the PMCA with

```
#pragma omp target map(tofrom: ctx, buf)
```

The PMCA now operates on the shared main memory and the caches of the host instead of its SPM. Thus, the next step is to allocate buffers in SPM and control the DMA engine inside the loop to transfer data while computations are running (third, orange bar). As the AES block cipher performs mostly byte-wise operations that map much worse to the basic RISC-V (`RV32IM`) ISA than to the more complex ARMv7-A ISA, performance is still relatively low. Since the full ISA of the PMCA is exposed, however, we can tune the code to leverage the extensions it offers: The RI5CY PEs [Gau+17] in this PMCA implement instructions that interpret the four bytes in a 32-bit word as elements of a byte vector, instructions to pack and unpack bytes from and into words, and instructions to shuffle and rotate bytes within a word. Such operations are at the heart of a block cipher, and using them massively improves performance (fourth, light green bar). We finally use the aforementioned `omp parallel for` annotation to parallelize encryption among 7 PEs, speeding up execution by 5.9× (fifth, dark green bar). With the PMCA running at 50 MHz on the FPGA, this is a modest 11 % faster than the dual-core ARM host. In a HeSoC IC, with host and PMCA running at 2 GHz and 800 MHz, respectively, the application would not saturate memory bandwidth, and offloading it to the PMCA could bring a speed-up of ca. 6.8×. Even more remarkably, the SVM-based offload is 90 % faster than a bare-metal implementation where all 8 PEs compute and DMA transfers run between SPM and physically-addressed shared memory (rightmost, gray bar). The reason is that when SVM is not used, the host must gather buffers from application virtual memory into a dedicated, physically-addressed, uncached memory region before the offload and scatter the buffers back after the offload, and the host is very inefficient at doing this.

In summary, these results show (i) that HEROv1 has the potential to effectively exploit both the standard ARM ISA and a specialized RISC-V ISA, bringing the acceleration potential of parallel, domain-specific PMCAs to bear, and (ii) that HEROv1's hardware and software enable host and PMCA to efficiently share data at a minimal programming effort and with a performance impact that ranges from slightly negative to significantly positive compared to copy-based memory sharing.

2.4 Related Work

HEROv1 extends the principle of prototyping computer architectures on FPGAs to HeSoCs. In the FPGA Architecture Model Execution (FAME) taxonomy [Tan+10a], HEROv1 is a Direct FAME system, meaning it implements the target architecture with a one-to-one correspondence in clock cycles on an FPGA. More sophisticated FAME levels decouple timing and functionality, exchange structural equivalence for modeling abstractions, and share FPGA resources in time between components of the target architecture to increase model flexibility and emulation throughput. An example of a sophisticated FAME system is RAMP Gold [Tan+10b], which is designed for the rapid early-design-space exploration of manycore systems. It is cycle-accurate and comparable in throughput to HEROv1, but requires the development of a behavioral model that is not directly used in the silicon implementation. In contrast to highly sophisticated FAME systems, HEROv1 is not designed for early-stage design explorations but for the evaluation, advancement, and extension of a proven PMCA template and for studying the integration of PMCAs in a HeSoC. By staying as close to the silicon implementation as possible, co-development and maintenance of separate models are avoided. Commercial Direct FAME systems, such as Cadence Palladium and MentorGraphics Veloce, are targeted at the verification of entire ICs. To reach the required capacity, they employ custom logic simulation engines and highly intrusive tracing systems in addition to FPGAs. They come with proprietary tools and protective licenses at very high costs, which bars the vast majority of the research community from using them.

The Flexible Architecture Research Machine (FARM) [Ogu+10] is a system for prototyping custom hardware implemented on an FPGA that is connected to an AMD multiprocessor. Both FARM and HEROv1 provide a cache-coherent link to the host processor and data transfer (or DMA) engines. While FARM leaves the task of implementing an accelerator from scratch and integrating it with the system to the researcher, HEROv1 comes with a RISC-V manycore implementation, a heterogeneous toolchain, and tools to allow efficient hardware and software research using standard benchmarks and real-world applications.

Intel uses FPGAs to prototype heterogeneous – in their definition two sets of cores of the same ISA but different power-performance design points – architectures [Wan+10; Chi+12]. They combine a Xeon [Wan+10] and an Atom [Chi+12] CPU with an FPGA on which they implement up to four "very old" [Chi+12] Pentium 4 cores. While an evaluation platform with a Xeon and an Atom CPU (both hard-macro) was shared with selected academic partners, the reconfigurable, FPGA-based prototypes remain restricted to Intel [Chi+12]. HEROv1, on the other hand, is openly available, implements a modern RISC-V manycore on an FPGA, and uses the extended concept of heterogeneity with different ISAs.

HEROv1 is more than a PMCA implemented on an FPGA, but its PMCA implementation is nonetheless related to the following recent works: OpenPiton [Bal+16] is the first open-source, multithreaded manycore processor and is available in FPGA implementations for prototyping. Our PMCA implementation on the FPGA differs from that of OpenPiton in two ways: First, our PMCA implements the RISC-V ISA, which has recently gained a lot of momentum. Second, it allows evaluation on the FPGA with more cores: we currently support up to 64 cores compared to OpenPiton's maximum of 4 cores (both on a Xilinx Virtex, albeit of different size). GRVI Phalanx [Gra16] is an array of clusters of RISC-V cores interconnected by a network on chip (NoC). Cores, clusters, and the NoC are optimized for FPGAs and utilize FPGA blocks very efficiently, allowing to implement hundreds of RV32I cores on a mid-range FPGA. While FPGAs are *the* design target of GRVI Phalanx, HEROv1 uses FPGAs as a prototyping target to support a wide range of implementation targets and architectural exploration. Moreover, GRVI Phalanx is programmed bare-metal, whereas the PMCA on HEROv1 comes with a runtime that supports well-established programming paradigms such as OpenMP including seamless accelerator integration. lowRISC [Bra+14] has a work-in-progress open-source SoC implementing the RISC-V ISA. Its goal is to lower the barrier of entry to producing custom silicon by establishing an ecosystem of IPs around RISC-V cores. In contrast, HEROv1 aims to facilitate exploration on all layers of software and hardware in HeSoCs by implementing a modifiable, working full-stack prototype accompanied by tools for validation and evaluation of novel concepts.

Computing acceleration through accelerator offloading started to gain more traction with the advent of programmable GPUs. Since a few years, nearly all GPU models – ranging from embedded, mid-end IP cores [Arm17b] up to high-end, data-center acceleration boards [Nvi17] – can serve as a target for offloading application kernels from the host. However, implementing and optimizing a heterogeneous application for GPU-based systems is not a trivial task. Typically, the offloadable kernels must be implemented in separate source files using lower-level programming languages such as OpenCL or CUDA and are compiled online before offloading. This not only requires special compilers decoupled from the host toolchain, but it also means more programming effort and prevents fine-grained kernel offloading. These problems can be somewhat alleviated by heterogeneous toolchains with open-source OpenMP [FBM18; Mar+16] or OpenACC [Rey+12] GPU front ends. However, such front ends can only generate intermediate code and still require to invoke proprietary GPU compilers. The ISA of the accelerator is closed and internal functions remain inaccessible for the developer. There is, e.g., no DMA engine exposed to overlap computations with data transfers. To achieve such behavior for hiding main memory latency, DMA transfers must either be explicitly emulated using regular loads and stores [FBM18], or the kernel must inherently offer very high degrees of data-level parallelism. In addition, GPU drivers and RTE libraries are completely closed for most devices [Arm17b; Nvi17]. In contrast, the host side of HEROv1 is partially and its PMCA is fully open-source starting from the RTE libraries down to the actual hardware. This gives the developer the possibility to optimally leverage the available hardware and exploit the full potential of the PMCA platform with a fully-integrated toolchain.

Heterogeneous compilers have also been implemented by others, both in research [CM17; CMB18; MCB16] and commercially [Int18]. [CM17] implemented an OpenMP plugin for GCC 5 to offload to OpenRISC-based PMCAs. While the ISA in that work is also exposed to the toolchain, OpenRISC was not designed with domain-specific extensions in mind. In contrast, HEROv1's software stack and toolchain are capable of fully leveraging custom extensions of the RISC-V ISA, as shown by our experimental evaluation. Intel is offering OpenMP-based programming of its Xeon Phi accelerators [Int18]. Although both the Intel host CPU and the Xeon Phi accelerator implement the x86 ISA, thy differ in (vendor-defined) extensions. While GCC can be used to offload to Xeon Phi, the offloaded LTO itself must be generated with the proprietary Intel C compiler to make use of all vector extensions. Similar to GPUs, the accelerator ISA is thus not directly accessible to developers, barring them from exploiting all accelerator features in their libraries. In HEROv1, in contrast, RISC-V-based PEs can be extended with domain-specific instructions fully exposed to library developers through an end-to-end open software stack.

2.5 Summary and Limitations

We presented the hardware and software architecture of our heterogeneous research platform, which unites a hard-macro ARM Cortex-A host processor with a modifiable RISC-V-based PMCA implemented on an FPGA. The software stack of the platform simplifes porting of standard benchmarks and real-world applications, thereby enabling system-level research and efficient data sharing between host and PMCA at a low programming effort. We presented a prototype implementation that allowed us to study the parallel speedup over multiple PMCA clusters, the performance of the virtual memory subsystem, and the speedup of offloading from the host to the PMCA.

This first version of our heterogeneous research platform has serious limitations, however: First, the on-chip network of the PMCA is composed of FPGA-proprietary IPs that cannot be modified or simple open-source IPs that are very limited in configurability (for instance, they have a fixed data width). This means it is not yet possible to construct heterogeneous on-chip networks that are found in state-of-the-art HeSoCs. Second, atomic memory operations, which are paramount for scaling concurrent algorithms to a high number of threads, are entirely missing. (Only the test-and-set operation, which has a consensus number of only two, is supported on the L1 memory of the PMCA.) Third, the DMA engine can only access virtual addresses that are guaranteed not to miss in the shared TLB (e.g., by locking TLB entries), and the performance of the virtual memory subsystem is limited by a single miss handling thread and lack of any prefetching. Fourth, the heterogeneous compiler enforces the same data model (and thus the same width of pointers and addresses) for host and PMCAs. Additionally, PMCA-custom instructions can be used by library developers but not directly by application programmers. Both are fundamental limitations of the used compiler that will require a profound change of technology.

Those limitations will be resolved in the subsequent chapters, after which we can present and evaluate a state-of-the-art research platform.

Chapter 3

High-Performance Non-Coherent On-Chip Communication

On-chip networks are the primary means of communication inside modern multi- and many-core processing SoCs [Jer+17; DT03; BD06; KC18]. As the number of cores, the heterogeneity of components, and the on- and off-chip bandwidth continue to grow to meet ever higher application demands, on-chip networks continue to gain importance. Decades of research on on-chip networks were instrumental for breakthroughs in scalability of homogeneous shared-memory multiprocessors, and a continuation of this research is necessary to realize the full potential of many-core accelerators and accelerator-rich heterogeneous SoCs.

Ideally, SoC designers could compose on-chip networks from a platform of components according to the requirements of their application. The central design goals of such a platform are:

(G1) Elementary, modular components that can implement any topology and that separate concerns such as routing and buffering.

(G2) Parametrizable components (e.g., data width, transaction concurrency) to cover a large design space.

(G3) Bridging components to connect heterogeneous SoC elements (e.g., GPU SMs, DMA engines, and domain-specific accelerators) and their subnetworks, each with unique, application-driven latency and bandwidth requirements.

(G4) Compliance with an industry-standard protocol for extensibility, third-party compatibility, and verifiability.

(G5) Detailed characterization of the complexity and trade-offs of the components in terms of performance vs. cost (area, power) to guide design and optimization efforts.

Commercial offerings that meet (parts of) these goals exist from multiple vendors (details in § 3.4), but their microarchitecture, complexity, and performance are well-guarded trade secrets. Research has also worked toward those goals (details in § 3.4), but, to the best of our knowledge, an end-to-end platform for non-coherent on-chip communication that meets the needs of heterogeneous SoCs has not been presented yet in open literature and is not available as open-source hardware.

In this chapter, we fill this gap with these contributions:

1. We present a modular, topology-agnostic **(G1)**, high-performance on-chip communication platform of parametrizable components **(G2)** for a state-of-the-art, industry-standard protocol **(G4)** (§ 3.1). The components include bridges and converters to link subnetworks with different bandwidth and concurrency properties **(G3)**. We publish the modules of our platform, implemented in industry-standard SystemVerilog, under a permissive open-source license for research and industrial usage.

2. We discuss microarchitectural trade-offs and timing/area characteristics of the modules in our platform **(G5)**, both theoretically/asymptotically and with topographical synthesis results (§ 3.2). We show that our modules can be composed to build high-bandwidth (e.g., 2.5 GHz and 1024 bit data width), end-to-end on-chip communication fabrics (e.g., DMA engine to memory controller), with high degrees of concurrency (e.g., up to 256 independent concurrent transactions) and flexibility (e.g., 64-bit subnetworks).

3. We design and implement (post-P&R) a state-of-the-art many-core machine learning training (MLT) accelerator in a modern 22 nm technology (§ 3.3), where our communication fabric scales to 1024 cores on a die, which deliver more than 2 Tdpflop/s, providing 32 TB/s cross-sectional bandwidth at only 24 ns round-trip latency between any two cores.

We focus on non-coherent on-chip communication for two main reasons: First, coherent on-chip communication in homogeneous many-core processors has been studied extensively (see § 3.4 for an overview). Second, many complex heterogeneous SoCs (e.g., mobile application SoCs [Qua20], high-speed networking SoCs [Whe19]) and massively parallel data processing architectures (e.g., general-purpose graphics processing units (GPGPUs) [Smi20]) are not or only partially cache-coherent.

This chapter is organized as follows: We present the architecture of our on-chip communication platform in § 3.1 and characterize its performance and complexity in § 3.2. We then use our platform to design, implement, and evaluate the communication fabric of a state-of-the-art many-core MLT accelerator in § 3.3. Finally, we compare with related work in § 3.4 and conclude in § 3.5.

3.1 Architecture

Current on-chip communication is centered around the premise of high-bandwidth point-to-point data transfers. To fulfill this premise despite increasing point-to-point latency, three central traits of current on-chip communication protocols are: *burst-based transactions*, *multiple outstanding transactions*, and *transaction reordering*. Our design targets these central traits in general, so the concepts we present potentially apply to a wide range of modern on-chip protocols. More tangibly, we adhere to the latest revision (5) of the AMBA AXI [AXI-F.b]. AXI is one of the industry-dominant protocols and the only protocol with an open, royalty-free specification and a widespread adoption in current systems designed by many different companies. Other protocols with similar properties are discussed in § 3.4.

Terminology and Protocol Essentials

A *module* is a distinct functional unit that has at least one on-chip network port. A *port* is a collection of input and output signals of a module. A port can be either a *master port*, on which the module initiates transactions, or a *slave port*, on which the module responds to transactions. One module can have multiple slave and master ports. We collectively call the five independently-handshaked channels connecting a master port to a slave port a *bundle*. Each *channel* consists of multiple isodirectional payload signals and two signals for bi-directional flow control. A *beat* is the data transferred on one channel upon one handshake; it is the smallest unit of communication. We focus on *valid-ready* flow control, where the channel master drives the *valid* signal and the payload signals and the channel slave drives the *ready* signal (but other flow control schemes, e.g., credit-based, are possible). A *handshake* occurs when *valid* and *ready* are high on a rising clock edge.

There are two essential rules in *valid-ready* flow control:

(F1) Stability Rule: Once *valid* is high, *valid* and the payload must not change until the handshake occurs.

(F2) Acyclicity Rule: The channel slave may depend on *valid* to be high before setting *ready* high, but the channel master may not depend on *ready* to be high before setting *valid* high.

Each transaction has a *direction* (read or write): A *write transaction* starts with one beat on the *write command channel* followed by one or multiple beats on the *write data channel* and ends with a single beat on the *write response channel*. A *read transaction* starts with one beat on the *read command channel* and ends with one or the last of multiple beats on the *read response channel*. A transaction is *outstanding* in the time interval starting with the handshake of the command beat and ending with the handshake of the (last) response beat. Each transaction has a numeric *ID*.

IDs define the order of transactions and beats according to the following rules:

(O1) Inter-Transaction Ordering: Any two transactions in the same direction and ID are ordered.

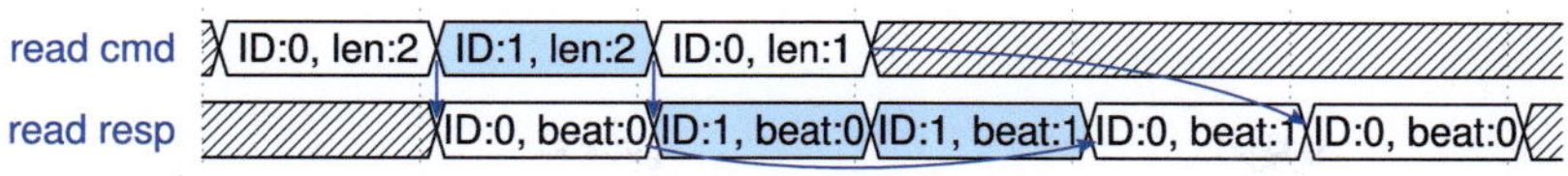

Figure 3.1: Transaction ID and ordering example. Three read commands are issued; the first and last have the same ID, the one between has a different ID. This situation might occur, e.g., if the transactions are issued by different masters. The first response beat is for the first command. After that follow both response beats for the second command. This interleaving of response beats is allowed because the first two commands have different IDs, so their responses may be interleaved. Then follows the second response beat for the first command. The response beat for the last command comes at the end, because it must not come before the last response beat of the first command (which has the same ID).

(O2) Response Ordering: Any two responses with the same direction and ID must be in the same order as their commands.

(O3) Write Beat Ordering: Write data beats do not have an ID and are therefore always ordered.

An example of IDs and their ordering is shown in Fig. 3.1.

An overview of the modules in our on-chip communication platform is given in Table 3.1. In this section, we discuss their microarchitecture and design trade-offs, from elementary components through all essential interconnecting modules to endpoints of increasing complexity.

3.1.1 Elementary Components: Network (De)Muxes

Our network multiplexers and demultiplexers are the elementary components that join multiple ports to one and split one port into multiple, respectively. In doing so, they must adhere to the relations between the channels and to the ordering rules **(O1–3)**. They are obviously used to build network junctions (e.g., crossbars), but they can be reused far beyond that because they implement a central part of the communication protocol. In fact, these elementary components are essential for almost all modules of our platform and can be used to design custom endpoints

Category	**Module**	**Section**
Elementary Components	Network Multiplexer	3.1.1
	Network Demultiplexer	3.1.1
Network Junctions	Crossbar	3.1.2
	Crosspoint	3.1.2
Concurrency Control	ID Remapper	3.1.3
	ID Serializer	3.1.3
Data Width Converters	Data Upsizer	3.1.4
	Data Downsizer	3.1.4
Clock Domain Crossing	Clock Domain Crossing	3.1.5
Data Movement	DMA Engine	3.1.6
On-Chip Memory Endpoints	Simplex Memory Controller	3.1.7
	Duplex Memory Controller	3.1.7
	Last Level Cache	3.1.8

Table 3.1: Overview of the modules in our on-chip communication platform.

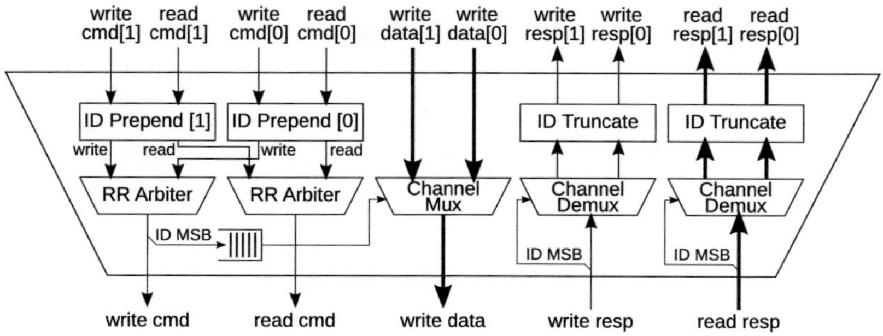

Figure 3.2: Architecture of our network **multiplexer**, drawn with two slave ports (at the top). Here and in Fig. 3.3, 'Channel (De)mux' refer to logic (de)multiplexers with additional handshake logic for one channel. In all later figures, 'Mux' refers to the entire module in Fig. 3.2 and 'Demux' refers to the entire module in Fig. 3.3.

without having to deal with all protocol intricacies. Each of our *network* (de)multiplexers contains simple *logic* (de)multiplexers, but it also contains other components to implement the protocol. In the remainder of this article, '(de)multiplexer' refers to a *network* (de)multiplexer unless explicitly preceded by 'logic'.

Network Multiplexer

The multiplexer, which connects multiple slave ports to one master port, consists of multiplexing components for the forward channels and demultiplexing components for the backward channels. The complexity lies in *de*multiplexing the backward channels, because the multiplexer needs the information to which output a beat on a backward channel must be routed. Multiplexing the command channels simply requires the selection of a *valid* beat, with the restriction that a selection must be stable once made **(F1)**.

Our multiplexer architecture is shown in Fig. 3.2. We first prepend the ID of each command beat with the number of the slave port. We then select among beats on the command channels with round-robin (RR) arbitration trees. For writes, the decision is forwarded through a first-in first-out buffer (FIFO) to a multiplexer for the write data beats, which is sufficient due to **(O3)**. As commands out of our multiplexer carry the input port information in the most significant bits (MSBs) of their ID, routing responses is as simple as demultiplexing based on the MSBs and then truncating the ID to the original width. Another key advantage is that transactions with the same ID from any two different slave ports remain independent, so **(O1)** does not restrict communication through our multiplexer. Note that *channel demultiplexing* means the payload is the same for all demux outputs and only the handshake signals are (de)multiplexed.

Alternative multiplexer architectures could do without extending the ID, for example by allowing only transactions with different IDs concurrently or by remapping IDs internally. However, the former restricts communication, and the latter significantly increases the complexity of the multiplexer. Nonetheless, some network modules grow exponentially in complexity with the ID width. We have a modular solution to this challenge with the ID width converters discussed in § 3.1.3.

Network Demultiplexer

The demultiplexer, which connects one slave port to multiple master ports, is more complex than the multiplexer due to the ordering rules: When the demultiplexer gets two commands with the same ID and direction **(O1)** that go to two different master ports, it must deliver the corresponding responses in the same order **(O2)**. After the demultiplexer, however, transactions on different master ports are independent, so the demultiplexer cannot rely on the order of downstream responses to fulfill **(O2)**.

Our demultiplexer architecture, shown in Fig. 3.3, solves this by enforcing that all concurrent transactions with the same direction and ID target the same master port. For example, when a write with ID A targets master port 0, it is only forwarded if no writes with ID A to master ports other than 0 have outstanding responses; otherwise, the write must wait. To track this information, the demultiplexer contains one counter and one index register per ID and direction. Commands that fulfill the aforementioned requirement increase the counter; the (last) response decreases the counter. A channel register between the write command channel and the demultiplexer of the write data channel stores

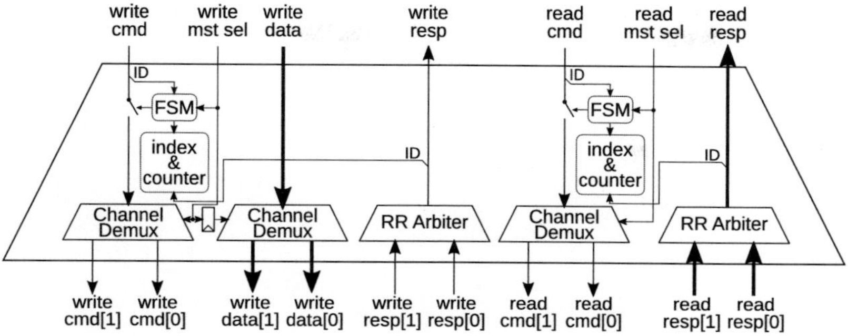

Figure 3.3: Architecture of our network **demultiplexer**, drawn with two master ports (at the bottom).

the master port index of an ongoing write burst while the command channel is independently handshaked **(F1)**. Write commands and data bursts are sent in lockstep due to **(O3)**; without this restriction, the write command and data channels could deadlock downstream. The multiple read and write response channels are joined through a round-robin arbitration tree.

Alternative demultiplexer architectures could do without requiring all concurrent transactions with the same direction and ID to target the same master port, for example by remapping IDs internally. However, this significantly increases the complexity of the demultiplexer, which would have to reorder responses internally to fulfill **(O2)**. Instead of introducing this complexity, we let a master use different IDs for different endpoints if it can handle out-of-order responses.

Compared with a 1-to-N crossbar, the demultiplexer has a fundamental advantage concerning how transactions are routed: With the crossbar, the *address* of a transaction determines to which master port it is routed. With the demultiplexer, the *select inputs* (one for reads, one for writes) determine to which master port a transaction is routed. This means a module instantiating the demultiplexer can freely decide which submodule handles a transaction. That decision does not even have to be based on the properties of the transaction but could, for example, be a function of the state of the module. This difference implies that the

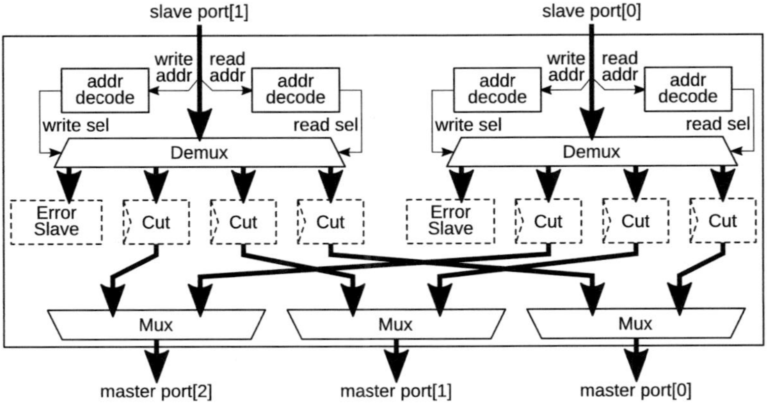

Figure 3.4: Architecture of our **crossbar**, drawn with two slave and three master ports. Each fat arrow represents a bundle, with the arrow head pointing in the direction of the command channels. Components with dashed outline are optional.

demultiplexer is a more universal elementary component than a 1-to-N crossbar.

Logic demultiplexers are so universally used in digital circuits that our *network* demultiplexer may seem like a trivial sequel. However, as the architecture depicted in Fig. 3.3 and described in this section shows, our demultiplexer handles crucial and complex parts of the protocol (**(O1–3)**, **(F1)**). Thus, even though our demultiplexer simply takes a select signal to route transactions, it unburdens user modules from dealing with intricacies of the protocol while it enables them to arbitrarily route transactions to submodules or ports.

3.1.2 Network Junctions: Crossbars and Crosspoints

Crossbar

The elementary components in § 3.1.1 can be combined to form a fully-connected crossbar, shown in Fig. 3.4, where each slave port has a dedicated connection to each master port.

At each slave port, two address decoders (one for reads, one for writes) drive the selection signals of a demultiplexer. In the standard configuration, all slave ports use the same addresses for one master port, but different configurations would be possible. There are two alternatives for handling transactions to an address that is not defined in a decoder. First, one master port can be defined as default port. This is useful, for example, in a hierarchical topology where each downlink has a specific range of addresses and any address outside the downlink addresses is sent to higher hierarchy levels through the uplink. Second, one can instantiate an error slave, which terminates all transactions with protocol-compliant error responses. These two alternatives can be selected per slave port with a synthesis parameter.

Optional pipeline registers can be inserted on all or some of the five channels of each internal bundle. These registers cut all combinational signals (including handshake signals), thereby adding a cycle of latency per channel and pipelining the crossbar so its critical path is no longer than that of the demultiplexer or multiplexer. These pipeline registers can be added without risking deadlocks, but this is not trivial: Of the four Coffman conditions [CES71], (1) Mutual Exclusion is fulfilled on the write data channel after the multiplexer, (2) Hold and Wait is fulfilled as each pipeline register must hold its value once filled, (3) No Preemption is fulfilled by **(O3)** on the write data channel, and (4) Circular Wait would be fulfilled by round-robin arbitration of write command and data beats. However, the demultiplexer breaks condition (4) by restricting write commands to be issued in lockstep with write data bursts (i.e., the next write command is only issued after the previous write data burst has completed), thereby preventing deadlocks despite pipeline registers, which introduce condition (2).

Crosspoint

As the multiplexers in the crossbar expand the ID width, the master ports of the crossbar have a wider ID than the slave ports. This prevents the direct use of our crossbar as nodes in a regular on-chip network where each node (also called "router" or "switch") has isomorphous slave and master ports. To solve this problem, we introduce a crosspoint.

Our crosspoint, shown in Fig. 3.5, has three additional properties over the crossbar that make it better suited for composing arbitary regular on-chip topologies. First, it contains a crossbar that is not necessarily fully connected: The connection between any slave and

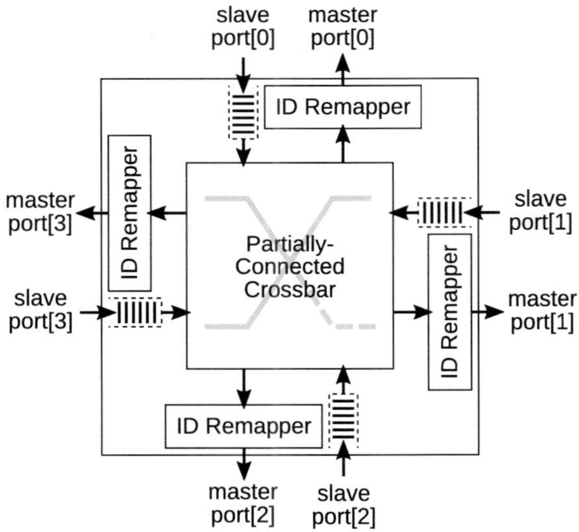

Figure 3.5: Architecture of our **crosspoint**, drawn with four slave and master ports. Each arrow represents a bundle, with the arrow head pointing in the direction of the command channels. The input queues are optional.

master port can be omitted with a synthesis parameter. This is useful to prevent routing loops when a module has both a master and a slave port into the crosspoint, and it minimizes the physical resources on links that would be unused. All flow and arbitration control logic of the crosspoint is inside the crossbar. Second, the crosspoint contains an ID remapper (§ 3.1.3) on each master port, which reduces the ID width to that of the slave ports. Thus, the slave and master ports of each crosspoint are isomorphous. Third, an input queue of configurable depth can be enabled for each slave port to reduce backpressure in mesh topologies.

3.1.3 Concurrent Transactions: ID Width Converters

The ID of transactions is central to their ordering **(O1–2)**. Essentially, the commands and responses of any two transactions can be independently reordered if they have different IDs. This makes a high number of possible IDs attractive to prevent bottlenecks due to ordering constraints. However, tracking a high number of IDs is complex for network components (e.g., demultiplexer § 3.1.1 and 3.2.1).

ID width converters are the on-chip network designer's instrument to balance the number of independent concurrent transactions vs. circuit complexity. We focus on *reducing* the ID width (as extending it is trivial). There are two first-order parameters for ID reduction: the width of IDs at the output, O, and the maximum number of unique IDs at the input, U. The relation between O and U determines whether all transactions that were independent at the input remain independent at the output: If $U \leq 2^O$, every unique ID at the input can be represented by a unique ID at the output, therefore retaining transaction independence. This means the sparsely used input ID space can be 'compressed' to a narrower, densely used output ID space by *remapping* IDs (§ 3.1.3). If $U > 2^O$, there are not enough output IDs to represent all U unique IDs. This means some transactions with originally different IDs will have to be mapped to the same ID, thereby *serializing* them (§ 3.1.3).

ID Remapper

Our ID remapper, shown in Fig. 3.6, remaps IDs with one table per direction. The table has as many entries as there are unique input IDs,

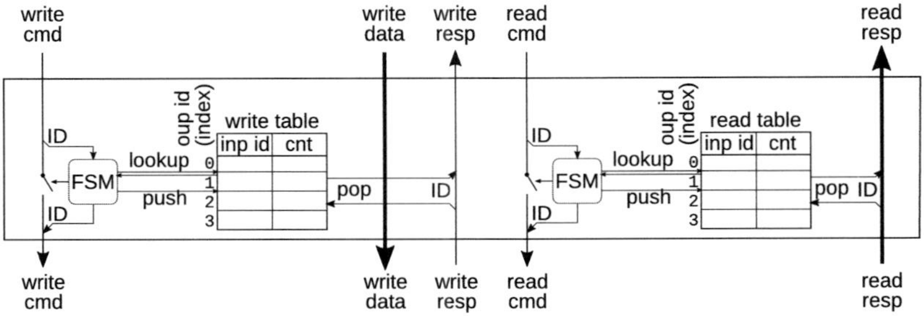

Figure 3.6: Architecture of our **ID remapper**, drawn with up to four unique concurrent IDs (per direction).

and it is indexed by the output ID. Each table entry has two fields: the input ID and a counter that records how many transactions with the same ID are in flight. The counter is incremented on command handshakes and decremented on (last) response handshakes. The mapping from input to output IDs is injective. Obtaining the input ID from an output ID (to remap responses) is as simple as indexing the table. Determining the output ID for an input ID (to remap commands) requires a comparison of the input ID to all IDs in the table. If the table currently contains an entry for the input ID, the same output ID must be used **(O1)**. If the table does not currently contain an entry for the input ID, the output ID is the index of the next free table entry.

Alternative ID remapper architectures could feature an additional table indexed by input IDs to look output IDs up. However, under the assumption of the remapper that the input ID space is sparse, such an additional table would be mostly empty. Therefore, it would be a poor usage of hardware resources and we omit it at the cost of a longer ID translation path, which could be pipelined.

ID Serializer

If the number of unique IDs at the input of the ID width converter, U, exceeds the number of available IDs at the output, 2^O, both the input and the output ID space are densely used. In this case, it is not possible to retain the uniqueness of all IDs during conversion, and we call the transformation that imposes additional ordering *serialization*.

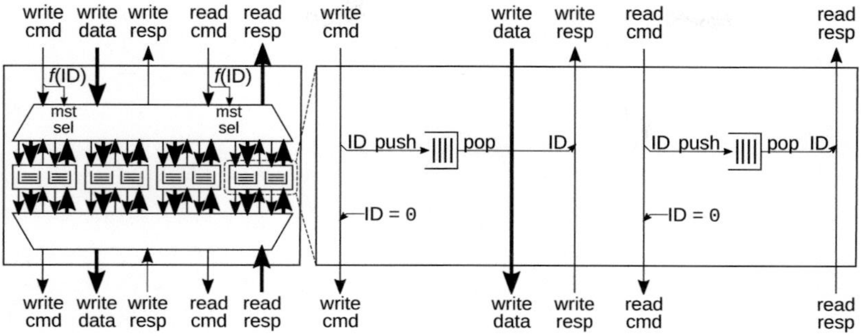

Figure 3.7: Architecture of our **ID serializer**, drawn with four master port IDs (per direction).

Serialized transactions still have concurrently outstanding commands, but they are now required to be handled in-order.

Our ID serializer, shown in Fig. 3.7, transforms IDs with one FIFO per direction and master port ID. At the slave port of the serializer, a demultiplexer assigns commands to one of the FIFO submodules through a combinational function f of the transaction ID (e.g., the ID modulo the number of master port IDs). The demultiplexer is a reduced configuration of our network demultiplexer (§ 3.1.1) without ID counters because f assigns identical IDs to the same master port (and thus the same output ID **(O1)**). In each FIFO submodule, the ID of a command is pushed into a FIFO and then truncated to zero. This FIFO reflects the transaction ID in responses **(O2)**, and the last response of a transaction pops from the FIFO. After the FIFOs, an instance of our network multiplexer (§ 3.1.1) assigns each transaction the index of its FIFO and merges the commands to the single master port of the ID serializer.

Alternative ID serializer architectures could use one memory where one linked list per master port ID is stored for ID reflection. This would allow to dynamically grow queues in memory rather than statically provisioning hardware resources to accommodate a fixed maximum of transactions per master port ID. However, pushing and popping IDs from this memory is on the critical path of the serializer, so we prefer the architecture with multiple FIFOs.

3.1.4 Data Width Converters

The data width of network components depends on their bandwidth requirements. For instance, the master port of a high-performance DMA engine might have 512 bit data width while that of a 64-bit processor core typically has 64 bit. This extends to subnetworks, e.g., separate networks for the DMA engine and the cores. However, as subnetworks with different data widths are joined, e.g., at endpoints such as memories, data width converters (DWCs) are required to convert between data widths. DWCs can be either *upsizers*, converting from narrow to wide, or *downsizers*, converting from wide to narrow. Although similar in purpose, up- and downsizer are not fully symmetric. In fact, the upsizer has higher performance requirements than the downsizer, since it must utilize the higher-bandwidth network as much as possible to minimize the impact on other components on the high-bandwidth network.

Data Upsizer

A data upsizer has a narrow slave port of data width D_{N} and a wider master port of data width D_{W}. In the simplest operating mode, pass-through, the upsizer does lane selection on read responses (Fig. 3.8a), selecting a slice of a wide incoming word, and lane steering on write data, aligning narrow incoming data into the wider outgoing word (Fig. 3.8b). In pass-through mode, the upsizer does not change the number of bytes transferred in each beat. This can be required by transaction attributes (e.g., to device memory). In terms of performance, however, this underutilizes the high-bandwidth network, which inherits the throughput of its low-bandwidth counterpart. Utilization can be increased by reshaping incoming bursts with many narrow beats into bursts with fewer wide beats: several narrow write data beats are packed into one wide beat, and one wide read response beat is serialized into several narrow beats.

Our data upsizer, shown in Fig. 3.8c, is capable of upsizing between interfaces of any data width. It is composed by two modules, read and write upsizers, that perform lane selection and steering, besides deciding whether to upsize the transaction based on its properties. Due to **(O3)**, only one write upsizer is needed, containing a buffer of width D_{W} to perform data packing. On the read response channel, the data upsizer handles a certain number of outstanding read transactions in parallel.

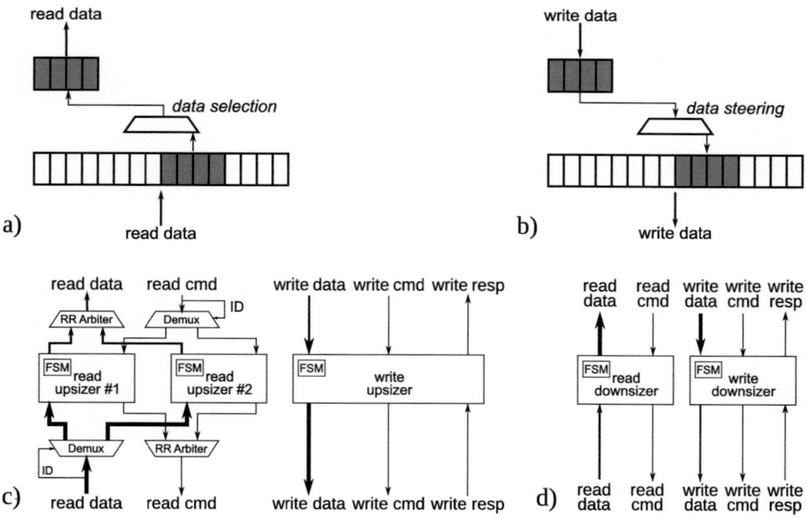

Figure 3.8: Architecture of our **data width converters (DWCs)**. (a) Data selection in the read response and (b) data steering in the write data channel of the upsizer. (c) Upsizer, drawn with two outstanding read transactions. (d) Downsizer.

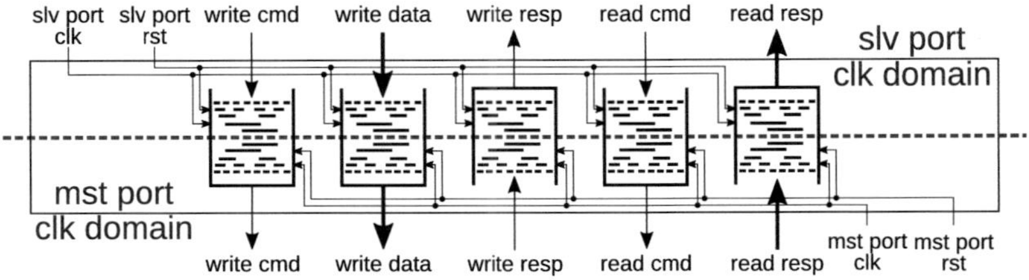

Figure 3.9: Architecture of our **clock domain crossing (CDC)**. Each channel goes through a CDC FIFO, which has two Gray-coded counters.

Each incoming read transaction is assigned an idle read upsizer, unless there is an active upsizer handling a transaction with the same ID. For that case, we ensure **(O1)** by enforcing that incoming transactions with the same ID are handled by the same read upsizer. Each read upsizer has a D_W buffer to hold incoming beats. This avoids blocking the wide read response channel during serialization.

Data Downsizer

A data downsizer has a wide slave port of data width D_W and a narrower master port of data width D_N. In the simplest operating mode, pass-through, the downsizer does steering on the read data channel and selection on the write data channel, symmetrical to the base operations of the data upsizer. Our downsizer, shown in Fig. 3.8d, differs from the upsizer in two key points: First, the downsizer has lower performance requirements than the data upsizer, since it connects to a lower-bandwidth subnetwork, e.g., peripherals. This means it does not need to support multiple outstanding reads. Second, when downsizing, the downsizer converts few wide beats into multiple narrow beats. It is possible that the resulting burst is longer than the longest buffer allowed by the protocol. In this case, the downsizer needs to break the incoming burst into a sequence of bursts. To handle this corner case, among others, the control logic of the read and write downsizers is more complex than those in the upsizer.

3.1.5 Clock Domain Crossing

A on-chip network can span multiple clock domains, yet all our modules have a single clock input[1] – except one: the clock domain crossing (CDC) has two clock inputs, one to which all signals of its slave port are synchronous and one for its master port. The CDC can be placed between any two modules in different clock domains. This enables the creation of independently-clocked subnetworks as well as the connection to endpoints that provide their own clock. In our CDC, shown in Fig. 3.9, each channel goes through a CDC FIFO, which has two Gray-coded counters: one for pushing the FIFO in one clock domain and one for popping from the FIFO in the other clock domain. The implementation follows well-established CDC principles [App+07; Cum08; SLB10].

3.1.6 Data Movement: DMA Engine

Transferring large amounts of data at high bandwidth requires dedicated components for data movement called *direct memory access (DMA) engines*. Our DMA architecture is designed to be modular, dividing the unit into two parts: a system-specific frontend and a backend implementing the data movement within the on-chip interconnect. We define a simple, yet well-defined interface uniting both parts: a one-dimensional and contiguous memory block of arbitrary length, source, and destination address, called *1D transfer*. We chose this interface abstraction because 1D transfers map very well to burst-based transactions. More complex transfers, such as multi-dimensional or strided accesses, are decomposed by the frontend into 1D transfers. As the frontend is highly system-specific, we will not discuss it.

In the backend, the *burst reshaper*, shown in Fig. 3.10a, divides the arbitrary-length 1D transfers into protocol-compliant bursts (adhering to, e.g., address boundaries and maximum number of beats). On arrival of a new 1D transfer, the burst converter loads length, source address, and destination address into internal registers. The *burst boundaries* process determines the number of bytes that can be requested in the next burst. With this, the burst reshaper calculates the address of

[1]For modules with a single clock input, we do not draw the clock input and clock wires to all sequential cells in the block diagram in order to not overcrowd the diagram.

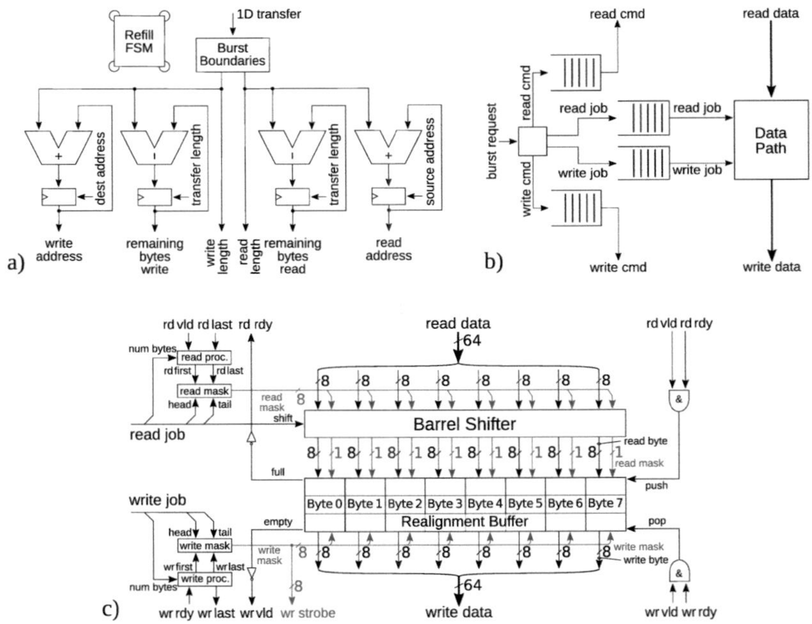

Figure 3.10: Architecture of our **DMA engine**. (a) Burst reshaper. (b) Data mover. (c) Data path, drawn for 64 bit data width.

the next burst and the remaining bytes left in the 1D transfer. Each protocol-compliant burst is then translated by the *data mover* unit, shown in Fig. 3.10b, into a read and a write command as well as a read and a write data job. The commands are issued as beats on the command channels. The data jobs are forwarded to the data path. The *data path*, shown in Fig. 3.10c, receives read data beats, realigns the data to compensate for different byte offsets between the read and write data streams, and issues write data beats. The data path consists of two independent processes. The read process is realigns and buffers incoming data. If a burst starts on an unaligned address, some leading bytes ("head") in the first beat are invalid and are masked. Similarly, a burst may end on an unaligned address, in which case some trailing bytes in the last beat ("tail") need to be masked. The write process drains data from the buffer as soon as it is available and masks it according to the destination address offset with the strobe signal of the write data channel.

3.1.7 On-Chip Memory Controllers

On-chip memories are an important class of endpoints for on-chip network transactions. In this section, we describe two memory controllers through which standard single-port static random access memory (SRAM) macros can be connected to the on-chip network.

Simplex Memory Controller

The architecture of our simplex on-chip memory controller is shown in Fig. 3.11. *Simplex* in this context means that the controller in each clock cycle can either read or write memory, as is natural for a single-port SRAM. The memory controller first translates read commands and write commands plus write data into memory commands. An arbiter then forwards either a read or a write memory command per clock cycle. This arbiter optionally takes quality of service (QoS) attributes of a command into account and can prioritize write beats, which cannot be interleaved due to **(O3)**, over read beats. A stream fork unit splits address and data, which go to the memory interface, and meta data (e.g., the transaction ID), which are used by the memory controller to form responses in the network protocol. A converter translates the

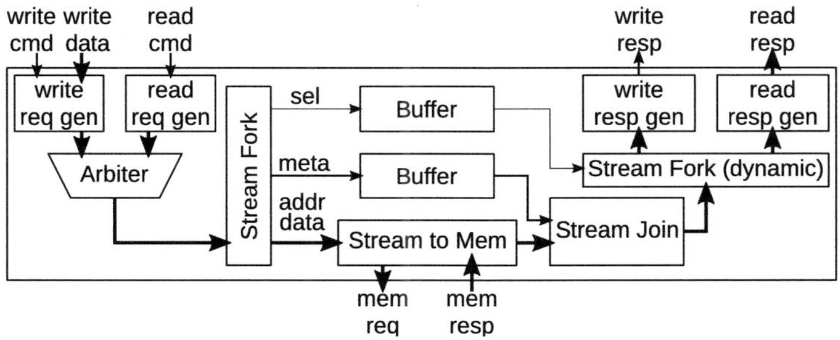

Figure 3.11: Architecture of our **simplex on-chip memory controller**, with the on-chip network slave port at the top and the memory master port at the bottom. The memory master port has the same data width as the network slave port.

address and data stream into memory interface signals (with stream flow control on the command and no handshaking on the response path). The memory responses are then joined with meta data to form read or write responses, which are finally issued on the corresponding network response channel.

The simplex memory controller cannot achieve the full bidirectional bandwidth of the duplex on-chip network interface, which has separate channels for read and write data. The duplex memory controller removes this limitation.

Duplex Memory Controller

The architecture of our duplex memory controller is shown in Fig. 3.12. To saturate the read and write data channels of the on-chip network simultaneously (thus *duplex*), this memory controller has at least two independent memory master ports as well as one simplex controller for writes and one for reads. A network demultiplexer statically routes all writes through the left controller and all reads through the right controller. The unused resources inside both simplex controllers are optimized away during synthesis. A logarithmic memory interconnect then routes each command to one of the memory master ports, which are address-interleaved.

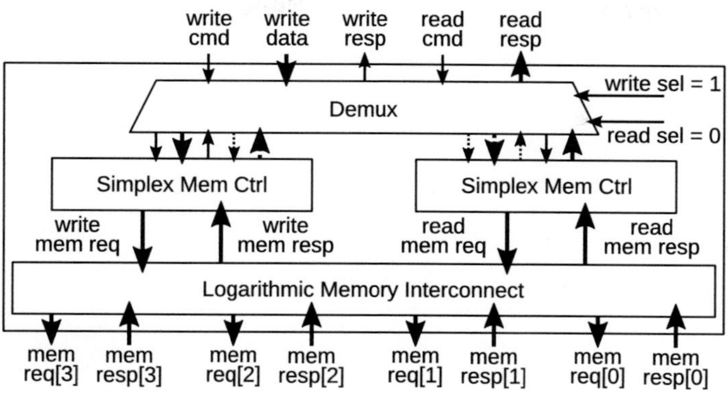

Figure 3.12: Architecture of our **duplex on-chip memory controller** with four address-interleaved memory master ports.

The duplex memory controller can fully saturate both the read and the write data channel of the on-chip network in the absence of conflicts on the memory ports. However, irregular traffic (e.g., misaligned addresses, mixed wide and narrow beats) can give rise to a significant conflict rate. To reduce conflicts, the *banking factor* (i.e., the number of memory master ports per network slave port) can be increased to any integer higher than 2 (at the cost of more wide and shallow SRAM macros when the memory capacity is to remain constant).

3.1.8 Last Level Cache

In contrast to the on-chip memory controllers of § 3.1.7, where the memory content is fully managed by software (so-called SPMs), a cache provides on-chip memory that is fully hardware-managed. As this chapter focuses on non-coherent on-chip communication, we present a non-coherent *last level cache (LLC)*. The purpose of this LLC is to reduce the latency and bandwidth between its slave (ingress) port and its master (refill) port. This is very useful, for example, in front of an off-chip memory controller, which has a much higher latency and lower bandwidth than on-chip memory.

Our LLC's set associativity, number of cache lines, and number of cache blocks per cache line are synthesis parameters, giving system

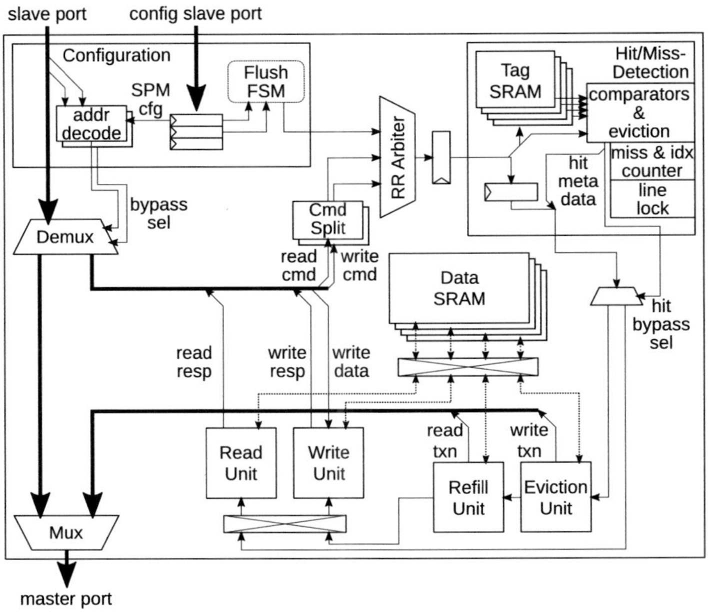

Figure 3.13: Architecture of our **last level cache (LLC)**.

designers complete control over the physical size and shape of the cache. The LLC uses a write-back, read-and-write-allocate data policy with pseudo-random eviction. The cache supports concurrent read and write accesses as well as eviction and refill operations. Reads are interleaved while adhering to **(O1–2)**. Transactions that hit in the cache can bypass earlier transactions that missed in the cache and are currently being serviced (i.e., eviction and refill) as far as permitted by **(O1)**.

As not all applications benefit from a hardware-managed cache, our LLC can be reconfigured at runtime to partially or fully become a software-managed SPM. This option is available at the granularity of individual cache sets. It is possible to use the entire data memory of our LLC as SPM. In that case, all accesses outside the address range of the SPM bypass the hit/miss logic of the LLC and are directly forwarded to the master port. This bypass is also used for non-cacheable transactions.

The architecture of our LLC is shown in Fig. 3.13. Like most components in our platform, the LLC is implemented with the stream-based control scheme that is natural to on-chip communication. The main idea is to start from the command and write data beats at the slave port, then transform, split, and merge them into descriptors that flow through the cache and give rise to new commands (for evictions and refills) and eventually to read and write responses. Starting at the slave port, commands are decoded by address and memory attributes and either sent to the bypass or into the core of the LLC. A command beat enters the cache over the command splitting units. These units split the command into descriptors, each of which targets exactly one cache line, and determine whether the access targets a cache set or an SPM region. Afterwards, the descriptors are arbitrated together with flush descriptors into a common pipeline. The descriptors then enter the hit-miss detection unit. Descriptors flagged as SPM simply flow through this unit, whereas all other descriptors perform a lookup inside the tag storage. The comparison and eviction unit determines the exact cache line and set of the descriptor. Additionally, this unit determines whether the descriptor gives rise to a refill or eviction. Descriptors that miss in the cache are sent to the eviction and refill pipeline, whereas descriptors that hit bypass this pipeline, which reduces their access latency.

Data consistency and ordering rules **(O1–3)** are enforced by two units: The index and miss counters prevent that a descriptor that hits overtakes another descriptor with the same ID that missed and is currently being served. The line lock allows only one descriptor to operate on a cache line and set at a time, which prevents data corruption that could occur from descriptors evicting a cache line used by another descriptor.

The data SRAMs are manipulated by four units: the eviction and refill units, which update the state of the data prior to a requested operation, and the read and write units, which perform the actual cache operation. All four units are connected over a logarithmic memory interconnect to the data SRAMs. The data width of the data channels and the SRAM data ports correspond to the cache block width. This setup allows all four units to concurrently have one descriptor each active on the data, thereby using the maximum available bandwidth of the slave and the master port of the LLC.

3.2 Implementation Results

This section provides quantitative and asymptotic complexity results for our network modules. These results are essential for architects to assess the feasibility and strike trade-offs in the design of on-chip networks. Our findings are summarized in § 3.2.9. Until there, this section discusses implementation results to derive the findings.

We implement the modules presented in § 3.1 in GlobalFoundries' 22 nm fully-depleted silicon-on-insulator (GF22FDX) technology, using a ten-metal stack and eight track SLVT/LVT flip-well standard cells characterized at typical conditions (0.8 V, 25 °C). We synthesize with Synopsys DesignCompiler 2019.12 using topographical mode, so physical place-and-route constraints, dimensions, and delays are taken into account. For the isolated implementation of the modules, each input is driven by a D-FF, and each output drives a D-FF. Unless we vary it in the evaluation, we set the address and data width to 64 bit and the slave port ID width to 6 bit. Before undergoing synthesis, all modules have been verified for protocol compliance in RTL simulation under extensive directed and constrained random verification tests.

3.2.1 Elementary Components: Network (De)Muxes

Network Multiplexer

The critical path of the multiplexer goes through from a slave port command channel through the arbitration tree on its handshake signals and the multiplexers on its payload signals to a master port command channel. For S slave ports, it scales with $\mathcal{O}(\log S)$ due to the logarithmic depth of the arbitration tree and the multiplexers. The area scales $\mathcal{O}(S)$ due to the linear area of the arbitration tree and the multiplexers. The area is further linear in the ID width and the maximum number of write transactions due to the FIFO between write command and data channel, but this part is usually negligible. Fig. 3.14 shows the area and timing characteristics of our multiplexer: for 2 to 32 slave ports, the critical path increases logarithmically from 190 to 270 ps, and the area increases linearly from 2 to 30 kGE.

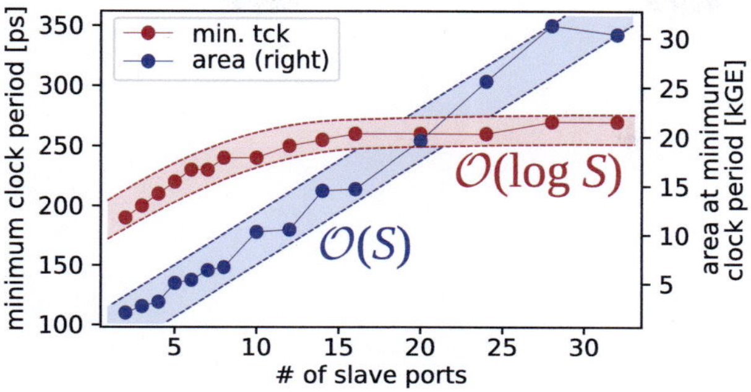

Figure 3.14: Minimum clock period and corresponding area of our network **multiplexer** in GF22FDX for 2 to 32 slave ports and 6 ID bits.

Network Demultiplexer

The critical path of the demultiplexer goes from a command channel at the slave port through ID lookup to a command channel on one of the master ports. It scales with $\mathcal{O}(M)$ as the channel demultiplexers grow linearly in area with the master ports and topographical synthesis takes the distance increase into account. The area scales with $\mathcal{O}(M)$ due to the linear area of the arbitration trees and the channel demultiplexers. The ID width I is critical for the demultiplexer: the area scales with $\mathcal{O}\left(2^I\right)$ due to the exponential number of counters (one for every possible ID), and the critical path scales with $\mathcal{O}(I)$ because every ID bit adds a multiplexer level in the indexing logic of the counters. Fig. 3.15 shows the area and timing characteristics of our demultiplexer: For 2 to 32 master ports and 6 ID bits (Fig. 3.15a), the critical path increases linearly from 330 to 430 ps, and the area increases linearly from 22 to 38 kGE. The curve is non-monotonic mainly in two points, where the synthesizer selects disproportionately strong and large buffers to reach the target frequency. For 4 master ports and 2 to 8 ID bits (Fig. 3.15b), the critical path increases linearly from 250 to 400 ps, and the area increases exponentially from 5 to 95 kGE. Depending on the ID width, the critical path can be significantly longer than in the multiplexer, so the demultiplexer will be the critical stage in a pipelined network junction.

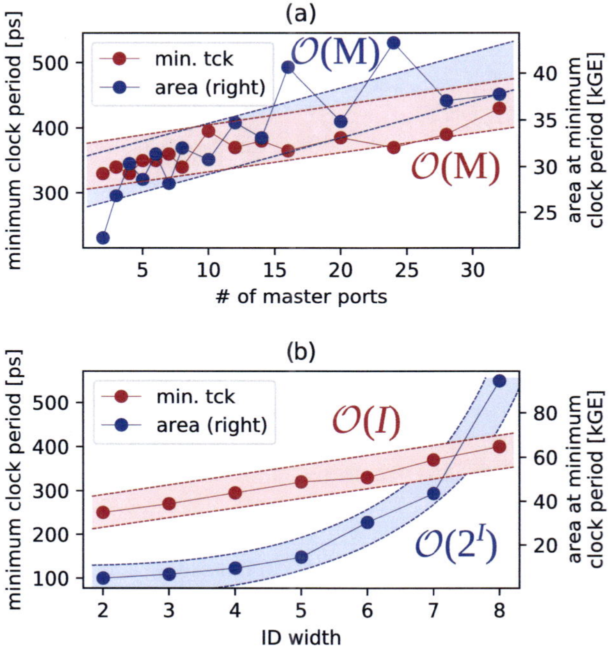

Figure 3.15: Minimum clock period and corresponding area of our network **demultiplexer** in GF22FDX: (a) with 2 to 32 master ports and 6 ID bits, and (b) with 4 master ports and 2 to 8 ID bits.

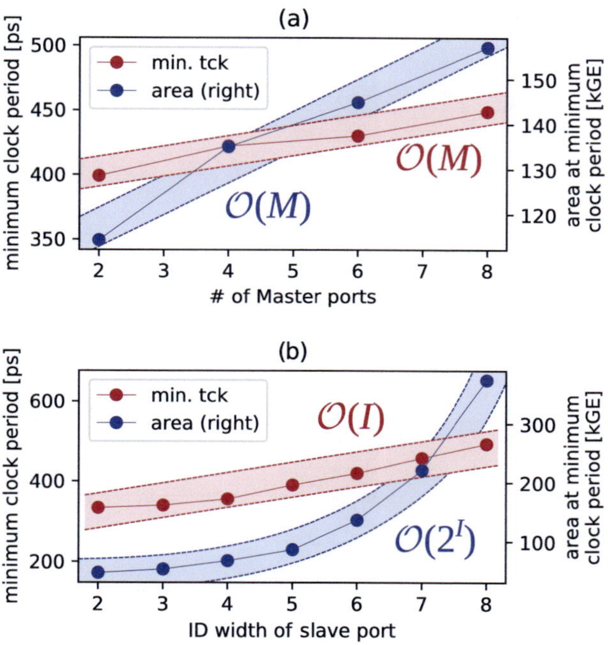

Figure 3.16: Minimum clock period and corresponding area of our **crossbar** with 4 slave ports, fully connected and unpipelined, in GF22FDX: (a) with 2 to 8 master ports, 4 slave ports and 6 ID bits, and (b) with 4 master ports and 2 to 8 ID bits at the slave port.

3.2.2 Network Junctions: Crossbars and Crosspoints

Crossbar

For a fully-connected crossbar with S slave ports, M master ports and I bits at the slave port, the critical path is dominated by the demultiplexer, thus scales with $\mathcal{O}(M+I)$. The area is the sum of the area of the S demultiplexers and M multiplexers plus a small overhead for each slave port for address decoding and the error slave (when instantiated). The area thus scales with $\mathcal{O}\left(MS+2^I S\right)$. Fig. 3.16 shows the area and timing characteristics of a fully-connected, unpipelined instance of our crossbar: For 4 slave ports, 2 to 8 master ports and 6 ID bits

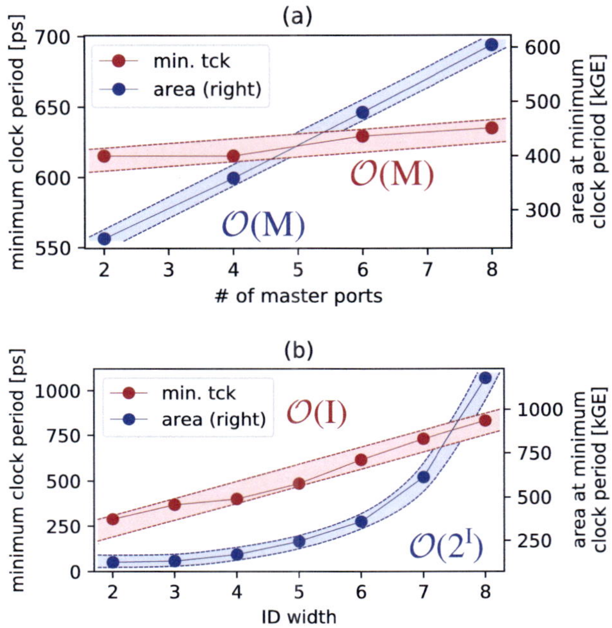

Figure 3.17: Minimum clock period and corresponding area of our **crosspoint** with 4 slave ports, fully connected and pipelined, in GF22FDX: (a) with 2 to 8 master ports, 4 slave ports and 6 ID bits, and (b) with 4 master ports and 2 to 8 ID bits at the ports.

(Fig. 3.16a), the critical path increases linearly from 400 to 450 ps, and the area increases linearly from 111 to 156 kGE. As was the case for the demultiplexer (§ 3.2.1), the ID width of the slave ports has significant impact on the critical path and area of the crossbar. For 4 master and 4 slave ports and 2 to 8 ID bits (Fig. 3.16b), the critical path increases linearly from 340 to 460 ps, and the area increases exponentially from 42 to 390 kGE.

Crosspoint

The critical path of a fully pipelined crosspoint goes from the internal pipeline register of a master port into the table of an ID remapper. For

M master ports (Fig. 3.17a), it scales with $\mathcal{O}(M)$ from 610 to 630 ns as topographical synthesis takes the area increase into account. The area also scales with $\mathcal{O}(M)$ but much more significantly from 243 to 587 kGE as the crossbar and the number of ID remappers scale linearly. Regarding the ID width I, the crosspoint is dominated by the demultiplexer: For 2 to 8 ID bits in a 4×4 configuration (Fig. 3.17b), the area scales with $\mathcal{O}(2^I)$ from 127 to 1181 kGE and the critical path scales with $\mathcal{O}(I)$ from 290 to 800 ps.

3.2.3 Concurrent Transactions: ID Width Converters

ID Remapper

The critical path of our ID remapper goes from the input ID through the ID equality comparators in in the table, through a leading-zero counter (LZC) to determine the matching or the first free output ID, into a table counter entry. For an input ID width of I, up to U concurrent unique IDs (per direction), and up to T transactions per ID, it scales with $\mathcal{O}(\log I + \log U + \log T)$. The area is dominated by the tables, which have U entries with $I + \log_2 T$ bit each. Additionally, the LZCs have an area of $\mathcal{O}(U \log U)$. The total area thus scales with $\mathcal{O}(U(I + \log T + \log U))$. Fig. 3.18 shows the area and timing characteristics of our ID remapper: For $U = 1$ to 64 concurrent unique IDs and $T = 8$ transactions per ID (Fig. 3.18a), the critical path increases logarithmically from 200 to 520 ps until $U = 48$ and then linearly to 640 ps for $U = 64$ as path delays due to the linearly growing table start to dominate. The area increases linearly from 1 to 41 kGE. The highest (rightmost) configuration can remap up to 512 transactions in both directions with up to 64 unique IDs concurrently, but the area and critical path costs are quite high. In comparison, for $U = 16$ concurrent unique IDs and $T = 1$ to 32 transactions per ID (Fig. 3.18b), the critical path increases logarithmically from 300 to 440 ps, and the area increases logarithmically from 7 to 16 kGE. Thus, the highest (rightmost) configuration can also remap up to 512 transactions but with only up to 16 unique IDs concurrently, at a $2.6\times$ lower area and $1.5\times$ shorter critical path.

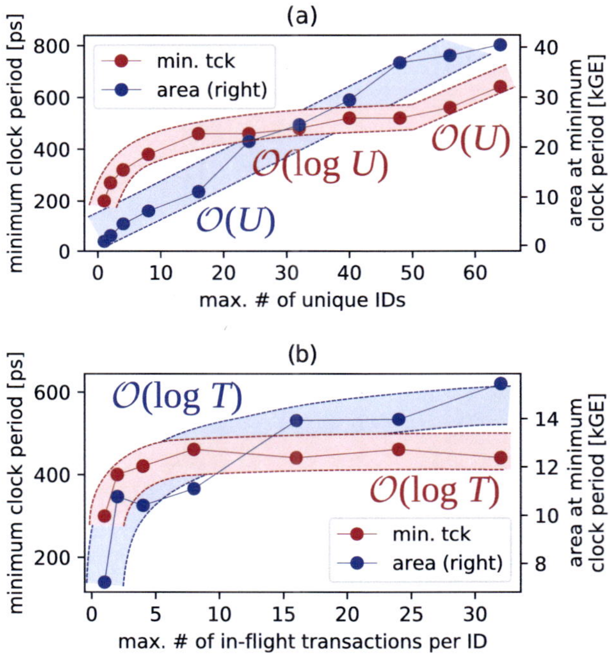

Figure 3.18: Minimum clock period and corresponding area of our **ID remapper** in GF22FDX: (a) for 1 to 64 concurrent unique IDs and 8 transactions per ID, and (b) for 16 concurrent unique IDs and 1 to 32 transactions per ID.

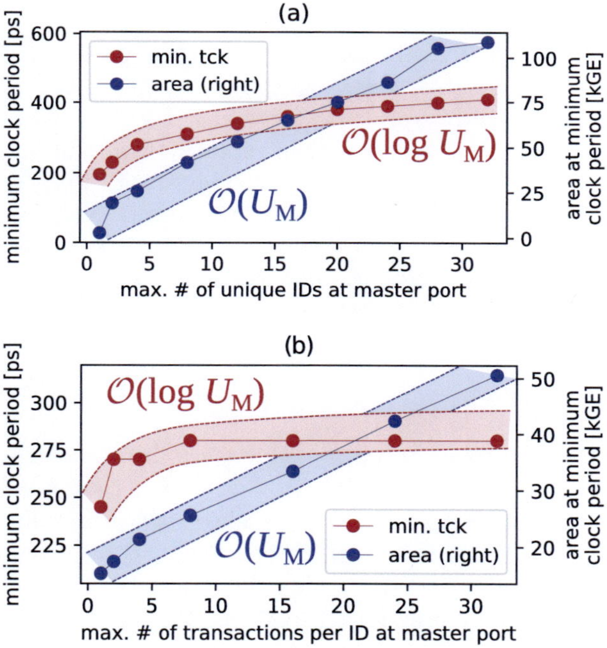

Figure 3.19: Minimum clock period and corresponding area of our **ID serializer** in GF22FDX: (a) for 1 to 32 IDs at the master port and 8 transactions per master port ID, and (b) for 4 IDs and 1 to 32 transactions per ID at the master port.

ID Serializer

The critical path of the ID serializer goes through the demultiplexer, the push side of the ID FIFO, and the arbitration tree in the multiplexer. For U_{M} IDs at the master port and T transactions per master port ID, it scales with $\mathcal{O}(\log U_{\mathrm{M}} + \log T)$. The area scales with $\mathcal{O}(U_{\mathrm{M}} + T)$ due to the linear area of all components in either U_{M} or T. Fig. 3.19 shows the area and timing characteristics of our serializer: For $U_{\mathrm{M}} = 1$ to 32 IDs at the master port and $T = 8$ transactions per master port ID (Fig. 3.19a), the critical path increases logarithmically from 195 to 410 ps, and the area increases linearly from 2 to 109 kGE. Clearly, compressing a densely used ID space is expensive in terms of area. This cost can be reduced by fixing U_{M} at a low value and varying T: For $U_{\mathrm{M}} = 4$ IDs and $T = 1$ to 32 transactions per ID at the master port (Fig. 3.19b), the critical path increases logarithmic from 245 to 280 ps, and the area increases linearly from 15 to 51 kGE. 128 concurrent transactions (in both directions) could therefore be serialized with $U_{\mathrm{M}} = 4, T = 32$ at 1.28× less area and 1.29× shorter critical path.

3.2.4 Data Width Converters

For our data downsizer between a wide slave port of width D_{W} and a narrow master port of width D_{N}, the critical path goes through the data selection and steering logic, scaling logarithmically with the downsize ratio $\mathcal{O}(\log(D_{\mathrm{W}}/D_{\mathrm{N}}))$. The area is $\mathcal{O}(D_{\mathrm{N}}D_{\mathrm{W}})$, the first term accounting for the multiplexing logic for data selection and steering, and the second accounting for the registers that hold a wide beat for data packing on the write data channel. Fig. 3.20a (left side) shows the area and timing characteristics of our downsizer: for a slave port of width 64 bits and a master port of width 8 to 32 bits, the critical path decreases with increasing width of the master port (and decreasing downsize ratio), from 390 to 365 ps, while the area grows linearly from 23 to 25 kGE.

For the data upsizer between a narrow slave port of width D_{N} and a wide master port of width D_{W}, the critical path goes through the data selection logic and the round-robin arbiter, scaling linearly with the number of read upsizers R and logarithmically with the upsize ratio, $\mathcal{O}(R \log(D_{\mathrm{W}}/D_{\mathrm{N}}))$. The area of the upsizer scales with $\mathcal{O}(RD_{\mathrm{N}}D_{\mathrm{W}})$, compounding the effect of the multiplexing logic for data selection and

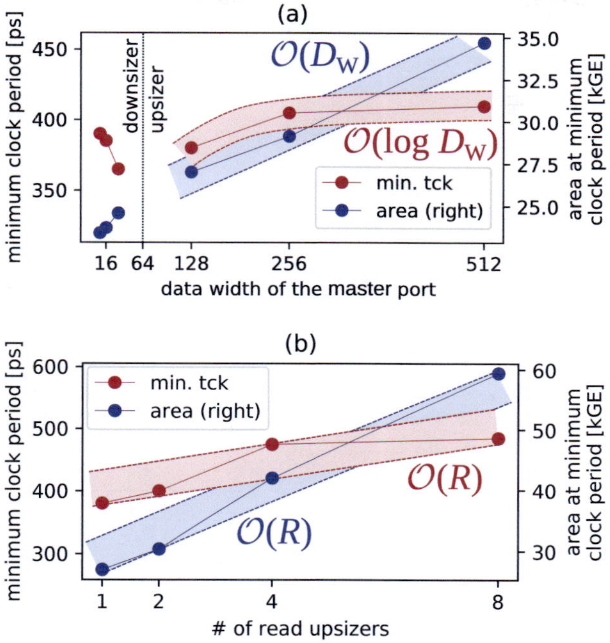

Figure 3.20: Minimum clock period and corresponding area of: (a) our **data downsizer and upsizer**, considering a slave port data width of 64 bit and a master port data width from 8 to 512 bit, and (b) our **data upsizer**, considering a slave port data width of 64 bit, a master port data width of 128 bit, and 1 to 8 read upsizers.

steering, D_N, and of the R D_W-bit registers holding wide beats for data serialization on the read data channel. Fig. 3.20a (right side) shows the area and timing characteristics of our upsizer: for a slave port of width 64 bits and a master port of width 128 to 512 bits, the critical path increases with the increasing upsize ratio, from 380 to 405 ps, while the area increases from 27 to 35 kGE. Fig. 3.20b shows the area and timing characteristics of the data upsizer from 64 to 128 bits, for 1 to 8 read upsizers. These have an important effect on the area and critical path of the upsizer. The critical path of the upsizer increases linearly from 380 to 485 ps, while the area increases from 27 to 59 kGE.

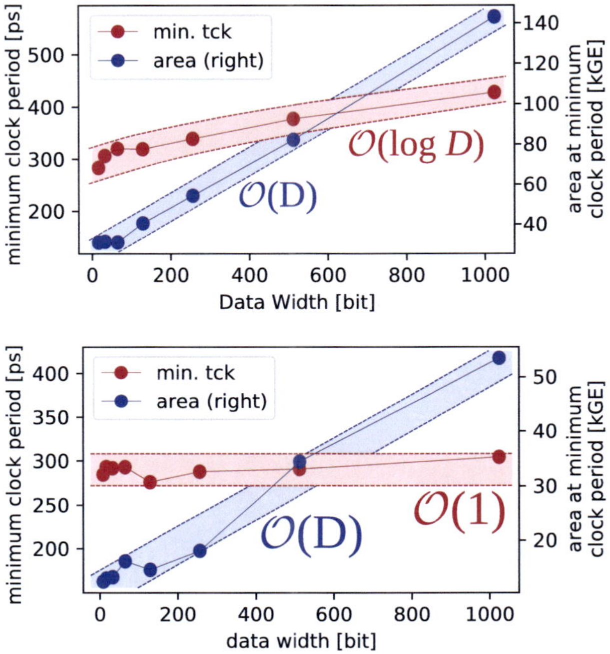

Figure 3.21: Minimum clock period and corresponding area in GF22FDX of (a) our **DMA engine** for 16 to 1024 bit data width, and (b) our **simplex on-chip memory controller** for 8 to 1024 bit data width.

3.2.5 Clock Domain Crossing

The area of the CDC scales linearly with the address, data, and ID widths. For 64 bit address and data width, 6 bit ID width, and a slave port clock frequency of 1 GHz, the CDC area is 27 kGE for master port clock frequencies from 0.1 to 2 GHz. Above that, the area increases exponentially but only up to 31 kGE for 5.5 GHz at the master port.

3.2.6 Data Streaming: DMA Engine

The area of the DMA engine scales with $\mathcal{O}(D)$, where D is the data width, due to the linearly growing alignment buffer. The critical path is dominated by the barrel shifter, which scales with $\mathcal{O}(\log D)$.

For 16 to 1024 bit data width (Fig. 3.21), the critical path increases logarithmically from 290 to 400 ps and the area increases linearly from 25 to 141 kGE. As the DMA engine uses the same ID for all transactions, the ID width affects neither area nor critical path.

3.2.7 On-Chip Memory Controllers

Simplex Memory Controller

For a simplex on-chip memory controller with a data width of D, the critical path is constant and found between the command slave channels and the memory master port. The critical path does not depend on D as the transformation of commands does not depend on the data width. Fig. 3.21b shows the area and timing characteristics: The area scales linearly with $\mathcal{O}(D)$ from 13 to 53 kGE; this linear dependency is caused by the dominant read response buffers needed for response path decoupling. The critical path remains roughly constant around 290 ps. The ID width has no impact on the critical path, as the simplex controller handles all commands in order and only buffers the ID for the response. The area scales with $\mathcal{O}(I)$ due to these buffers.

Duplex Memory Controller

The critical path of the duplex controller goes from the slave port command channels through the demultiplexer, one simplex memory controller, and the logarithmic memory interconnect to a memory command port. For a data width of D and B memory master ports, it scales with $\mathcal{O}(\log D)$. The area is composed of the demultiplexer, the two simplex memory controllers, and the logarithmic interconnect, and thus scales with $\mathcal{O}(B + D)$. Fig. 3.22 shows the area and timing characteristics of our duplex memory controller: For $D = 8$ to 1024 bit data width and $B = 2$ memory master ports (Fig. 3.22a), the critical path increases logarithmically from 280 to 330 ps, and the area increases linearly from 20 to 175 kGE. For $D = 64$ bit data width and $B = 2$ to 8 memory master ports (Fig. 3.22b), the critical path stays constant around 300 ps and the area scales with $\mathcal{O}(B)$ from 28 to 34 kGE. Regarding the ID width I, the complexity is defined by the demultiplexer.

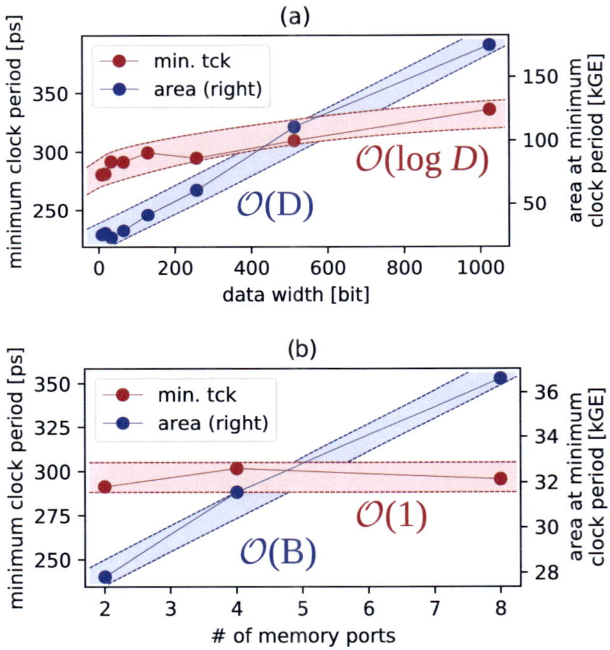

Figure 3.22: Minimum clock period and corresponding area of our **duplex on-chip memory controller** in GF22FDX: (a) for 8 to 1024 bit data width and two memory master ports, and (b) for 64 bit data width and 1 to 8 memory master ports.

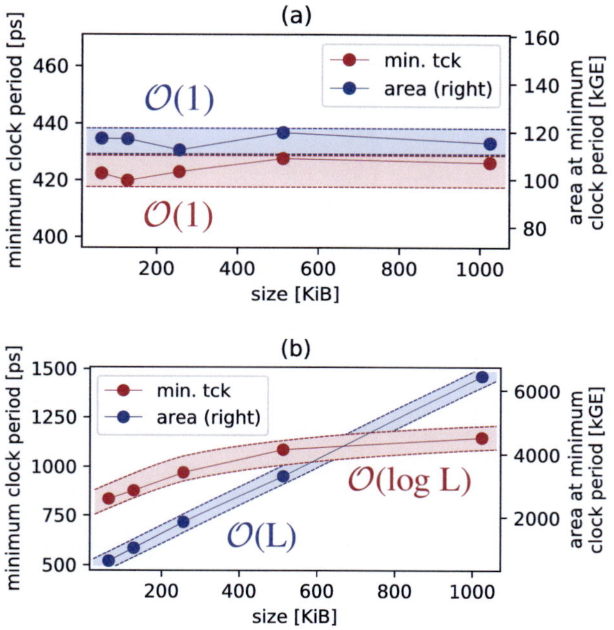

Figure 3.23: Minimum clock period and corresponding area of our **last level cache** in GF22FDX, with a set associativity of 4, 16 blocks per cache line, 8 B per block, and 64 bit addresses, (a) evaluated in isolation without SRAM and (b) evaluated together with SRAM.

3.2.8 Last Level Cache

We evaluate our LLC with a set associativity of 4, 16 blocks per cache line, and 8 B per block, and we vary the cache size through the number of cache lines L. Area and critical path of a cache are commonly dominated by its SRAM macros, but it is essential that the control logic adds only minimal overhead.

The control logic remains constant in area when increasing the cache size with L, as shown in Fig. 3.23a. Changing the ID width I would scale the area with $\mathcal{O}\left(2^I\right)$ due to the ID counters instantiated in the bypass multiplexer and the counters in the hit-miss unit. However, this is a secondary effect as those counters are relatively small compared to

	Critical Path	**Area**
Multiplexer	$\mathcal{O}(\log S)$	$\mathcal{O}(S)$
Demultiplexer	$\mathcal{O}(M+I)$	$\mathcal{O}(M+2^I)$
Crossbar	$\mathcal{O}(M+I)$	$\mathcal{O}(MS+2^I S)$
Crosspoint	$\mathcal{O}(M+I)$	$\mathcal{O}(M+2^I)$
ID Remapper	$\mathcal{O}(\log I+\log U+\log T)$	$\mathcal{O}(U(I+\log T+\log U))$
ID Serializer	$\mathcal{O}(\log U_{\mathrm{M}}+\log T)$	$\mathcal{O}(U_{\mathrm{M}}+T)$
Data Upsizer	$\mathcal{O}(R\log(D_{\mathrm{W}}/D_{\mathrm{N}}))$	$\mathcal{O}(RD_{\mathrm{W}}D_{\mathrm{N}})$
Data Downsizer	$\mathcal{O}(\log(D_{\mathrm{W}}/D_{\mathrm{N}}))$	$\mathcal{O}(D_{\mathrm{W}}D_{\mathrm{N}})$
DMA Engine	$\mathcal{O}(\log D)$	$\mathcal{O}(D)$
Simplex Mem. Ctrl.	$\mathcal{O}(1)$	$\mathcal{O}(D)$
Duplex Mem. Ctrl.	$\mathcal{O}(\log D+\log B+I)$	$\mathcal{O}(D+B+2^I)$
Last Level Cache	$\mathcal{O}(\log L)$	$\mathcal{O}(L+2^I)$

Legend: M = number of master ports; S = number of slave ports. D = data width; D_{W} = data width of the wide interface; D_{N} = data width of the narrow interface; I = ID width; U = concurrent unique IDs; U_{M} = concurrent unique IDs at the master port; T = concurrent transactions per ID. B = number of memory master ports. R = number of read upsizers. L = number of cache lines.

Table 3.2: Overview of the complexity of our network modules.

the remainder of the cache. The critical path is inside the tag lookup unit, starting at the tag memory, going through the tag comparators, and ending again in the tag memory. No logic on the critical path of the control unit changes with the number of cache lines L, therefore it is constant in L.

The LLC including the SRAM macros is characterized in Fig. 3.23b. Compared to the area of the control logic alone (Fig. 3.23a), the SRAM macros occupy 8 to 64 times more area for a cache size of 64 to 1024 KiB. Thus, the area occupied for control logic is below 10 % already at 128 KiB, and becomes marginal at larger sizes. The delay of the tag memory dominates the critical path of the design: as it increases logarithmically with the number of cache lines L, so does the critical path of the LLC.

3.2.9 Complexity Overview and Summary

An overview of the asymptotic complexity of our network modules is shown in Table 3.2. The critical path of all modules scales at worst linearly in their parameters, for most modules and parameters even logarithmically. As the absolute results of the minimum clock period show, the critical path of all modules remains below 500 ps post-topographical-synthesis in the large design space we evaluated. This shows our modules are suited for a wide range of target frequencies and bandwidths, up to 2 GHz. When even higher frequencies are required, most modules can be parametrized to have a critical path below 330 ps, which would allow to clock them up to 3 GHz. The area of most modules scales linearly in their parameters, with the notable exception of the ID width, which causes an exponential growth of the demultiplexer and all modules containing it. As the absolute results show, most modules fit a few tens of kGE when not pushed to the highest possible clock frequency and parametrization. Even more complex modules, such as a 4×4 crossbar with up to 256 independent concurrent transactions, fit in a modest 100 kGE when clocked at 2.5 GHz.

Power is another important aspect of on-chip networks. This chapter focuses on architecture, performance, and area, but our platform supports all state-of-the-art power optimization techniques (e.g., architectural power gating). Even for complex and high-performance instances such as the mentioned 100 kGE crossbar, the power consumption is in the order of just 35 mW under full load at 2.5 GHz, which would typically be less than 10 % of the power consumed by processor cores operating at a similar frequency; for instance, an ARM Cortex-A72 in a 16 nm FinFET technology at 2.5 GHz consumes around 1 W [Fru16].

While module-wise results are important to show the complexity and trade-offs in the microarchitecture of our on-chip communication platform, they of course cannot show the full picture of a real on-chip network. In the next section, we analyze a full on-chip network.

3.3 Full-System Case Study

In this section, we design, implement, and evaluate the on-chip network of a many-core floating-point accelerator, using the modules presented in

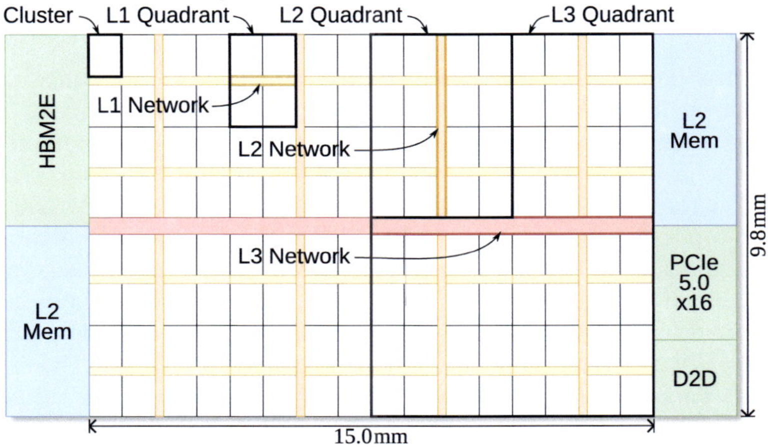

Figure 3.24: Conceptual floorplan of one Manticore [ZSB20] chiplet die.

this chapter. We use the technology and synthesis flow described in § 3.2 and additionally implement the networks with Cadence Innovus 19.1.

The *Manticore* architecture [ZSB20] is a state-of-the-art manycore processor for high-performance, high-efficiency, data-parallel floating-point computing. A Manticore accelerator consists of four chiplet dies on an interposer. Each chiplet, shown in Fig. 3.24, contains 1024 cores grouped in 128 clusters, one 8 GiB HBM2E controller and PHY, 27 MiB L2 memory, one Peripheral Component Interconnect Express (PCIe) 5.0 x16 controller and PHY, and three die-to-die link (D2D) PHYs to the other chiplets. Each cluster contains eight small 32-bit integer RISC-V cores, each controlling a large double-precision FPU, and 128 KiB L1 memory organized in 32 SRAM banks. As primary means for moving data into and out of L1, each cluster contains two of our DMA engines (§ 3.1.6, one for reads and one for writes), which are attached to the L1 memory and control a 512-bit-wide master port. DMA engines in other clusters can access the L1 memory through an additional 512-bit-wide slave port. Each cluster has a 64-bit master port to let its cores access external memory and a 64-bit slave port to let cores in other clusters access its L1 memory. Four clusters form an L1 quadrant, four L1 quadrants form an L2 quadrant, four L2 quadrants

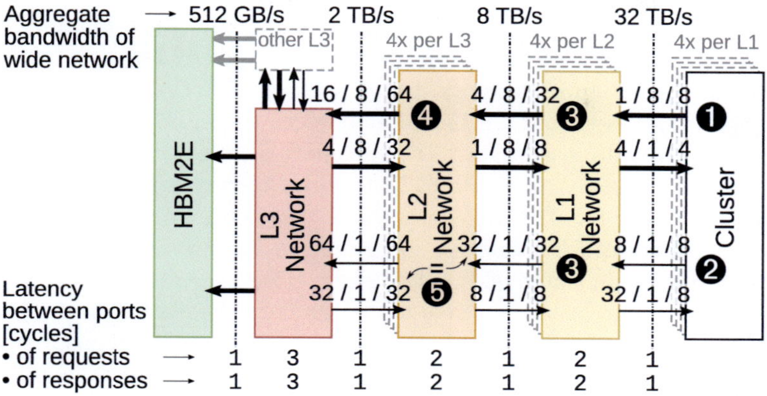

Figure 3.25: Manticore's on-chip network. Each arrow represents a bundle, with the arrow head pointing in the direction of the command channels. Fat arrows mean 512 bit data width, thin arrows 64 bit. Numbers above arrows indicate maximum transaction concurrency in the form *unique IDs / transactions per ID / total transactions per bundle* (reads and writes separate).

form an L3 quadrant, and two L3 quadrants form a chiplet. Manticore has been introduced in [ZSB20] without disclosing its on-chip network. In the remainder of this section, we describe the design, implementation, and performance of Manticore's on-chip network.

3.3.1 Network Design

Manticore's on-chip network is designed with four main goals: **(D1)** High bandwidth between units within the same quadrant for effective local data sharing. **(D2)** High bandwidth between the chiplet-level I/Os (i.e., HBM2E, PCIe, D2D) and any cluster for effective data input and output. **(D3)** Low latency between any two cores for efficient concurrency. **(D4)** Minimal interference between the wide bursts of the DMA engine and the word-wise accesses of the cores for maximum network utilization. The network, shown in Fig. 3.25, has the following properties to meet these goals: (1) Physically separate networks for traffic by DMA engines and cores to meet **(D4)**. (2) Tree topology to meet **(D2–3)**. (3) Fully-connected crossbars within each quadrant to meet **(D1)**. (4) The same data width and frequency for each bundle throughout the DMA network, to meet **(D2)**. The clock frequency of the entire network and all its endpoints (i.e., cores, DMA engines, L2 memory controllers, and HBM2E controller) is 1 GHz. The data width of the DMA network is set to 512 bit, which corresponds to one of the four ports into the HBM2E controller. Therefore, saturating the full HBM2E bandwidth requires concurrent transactions from only four DMA engines in different L2 quadrants. The data width of the core network is set to 64 bit, which is native for the load/store unit of a core.

The concurrency of transactions is another important aspect of the network design. The numbers above an arrow in Fig. 3.25 define the number of concurrent unique IDs, transactions per ID, and total transactions per bundle (reads and writes separate), respectively. ID width converters are placed in the network where required to reduce the ID width. Starting at the cluster, each DMA engine is in-order (thus has a single ID) and can have up to 8 outstanding transactions ➊. Transactions by the 8 cores in the cluster are independent, and each core can have at most 1 outstanding transaction ➋. The L1 network maintains the independence of all DMA and core transactions, and the number of unique IDs expands accordingly, as do the total transactions ➌. The L2 network maintains the independence of DMA transactions but limits their total below the sum of the incoming ports with ID remappers ➍. The reason is that the maximum roundtrip latency at this level is 60 cycles, so a higher number of concurrent transactions would not increase bandwidth or utilization. The concurrency on downlinks is generally constrained with ID remappers to match that of an uplink into the lower network, e.g., ➎. This means each network level can handle transactions from the uplink slave port in the same way as transactions from downlink slave ports.

3.3.2 Network Microarchitecture and Implementation

The microarchitecture and physical dimensions of one L1 and L3 network are shown in Fig. 3.26. (The L2 network is very similar to the L1 and omitted for brevity.) For the L1 network, the downlink ports are in the left third of each cluster, close to the cluster's memory and internal

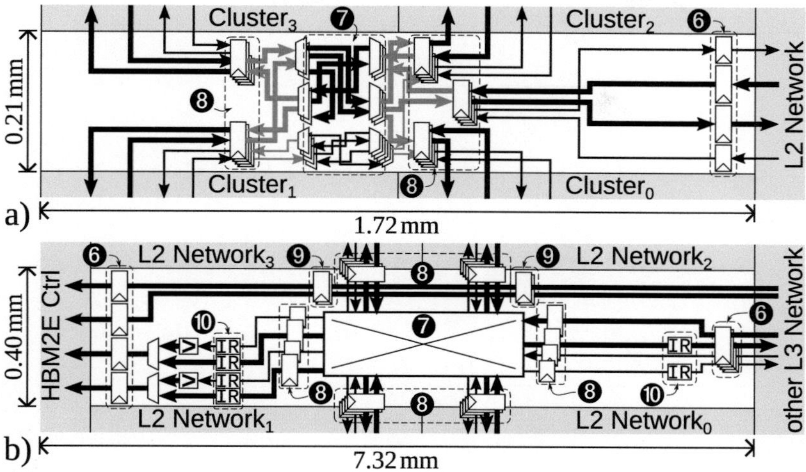

Figure 3.26: Microarchitecture and dimensions of Manticore's on-chip network. (a) L1 network. (b) L3 network. Only select connections are drawn for reasons of lucidity.

interconnect, and the uplink port is in the middle of the narrow side. For the L2 and L3 network, the downlink ports are at one quarter of the wide side (determined by the lower network level), and the uplink port is in the middle of the narrow side. To isolate the timing closure of individual network levels, we cut all paths at the uplink ports ❻. Correspondingly, all downlink inputs are driven by FFs and all downlink outputs drive FFs. There are two central challenges in the physical implementation of the networks. First, the extremely wide aspect ratio: while one wide dimension is determined by the side length quadrant, the other dimension should be as narrow as possible to minimize the area of the network. Second, routing and wire congestion: each of the five interfaces has ca. 3300 separate wires, and each network level is fully connected. Routing the wires of a single interface horizontally occupies a height of ca. 100 µm on all three metal layers available for inter-cell horizontal routing. To mitigate congestion, the crossbar, with its fanout of wires between demultiplexers and multiplexers, should be placed and routed as compact as possible ❼. The crossbar nonetheless incurs a

	Unit	L1	L2	L3	Entire Network
Clock Frequency	[GHz]	1.00	1.00	1.00	1.00
Routing Density*	[%]	59.6	49.6	45.7	—
Area per Inst.	[mm^2]	0.41	1.40	2.99	30.43
Power per Inst.	[mW]	8.1	12.8	17.2	396.0
# Insts. per Chiplet		32	8	2	1
Area per Chiplet	[mm^2]	13.21	11.23	5.98	30.43
Area per Chiplet†	[%]	9.05	7.69	4.10	20.84
Area per Core+FPU	[µm^2]	12 900	10 970	5840	29 710
Power per Chiplet	[mW]	259.2	102.4	34.4	396.0

*Routing density along wider dimension (i.e., where routing is denser).
†Relative to chiplet area without I/O controllers and PHYs.

Table 3.3: Implementation results of Manticore's on-chip network.

significant combinational delay. To accommodate this despite the long distances due to the extreme aspect ratio, we insert registers around the crossbar ❽. In contrast to pipelining inside the crossbar, much fewer registers are required, which again benefits the compact layout of the crossbar. In the L3 network (Fig. 3.26b), pairs of L2 networks share one port on the HBM2E controller. Cores on the narrow network access the wide HBM ports through data width converters. Because the HBM2E controller is located on the left side of the chiplet, the left L3 network simply feeds two connections from the right L3 network through pipeline registers to the controller ❾. ID remappers are used to reduce ID widths according to the concurrency design ❿.

The implementation results of Manticore's on-chip network are listed in Table 3.3. We have been able to close timing and design rule checking (DRC) of the entire network after place and route at 1 GHz. For this, we first loosely constrained the narrow dimension to determine the required number of pipeline registers around the crossbar, then we reduced the narrow dimension until the design could no longer be routed without failing timing or DRC. As the high routing densities show, the area of each network level is mainly determined by the available routing channels. The total area of the network is 30.43 mm^2, which

Figure	Unit	Convolution base	Convolution stacked	Convolution pipe'd	Fully Connected
Op. Intensity	[dpflop/B]	2.2	15.9	15.9	7.9
HBM BW	[GB/s]	262*	98	6	222
L3 Agg. BW	[GB/s]	262	98	6	222
L2 Agg. BW	[GB/s]	262	98	25	222
L1 Agg. BW	[GB/s]	262	98	98	222
Performance	[Gdpflop/s]	571	1638†	1638†	1638†

*Of which 256 GB/s are on the read channel, which is its maximum.
†This corresponds to an FPU utilization of ca. 80 %, which is the maximum all 8 FPUs in a cluster can sustain for real kernels.

Table 3.4: Performance of Manticore for different NN layer implementations.

is 20.84 % of the chiplet area without I/O controllers and PHYs. Put differently, Manticore's entire high-bandwidth, low-latency, hierarchical on-chip network requires 29 710 μm^2 per core. This is merely about the same area as one core (without any cache) and FPU, which are highly area-efficient. The total power of the network is 396 mW, which means only 0.4 mW per core (or less than 5 % extra power per core).

3.3.3 Network Performance

We characterize the performance of two fundamental neural network (NN) layers based on RTL simulations and analytical calculations, with a focus on the impact of the on-chip network. The two layers, a convolutional layer and a fully-connected layer, together account for 95 to 99 % of the floating-point operations (FLOPs) in MLT. The following is a condensed description of the NN implementation on Manticore, a comprehensive description is available as online supplement [2].

A convolutional NN layer transforms a 3D input volume (aka "feature map") to a 3D output volume through convolutions with filter kernels. More precisely, each input volume with dimensions $W_I \times W_I \times D_I$ is padded with P zeros in the spatial dimensions (resulting in $(W_I + 2P) \times$

[2] `https://arxiv.org/pdf/2104.08009.pdf`

$(W_I + 2P)$) and then convolved with K filter kernels with dimension $F \times F \times D_I$ in a stride S to produce an output volume with dimensions $W_O \times W_O \times D_O$, where $W_O = (W_I + 2P - F)/S + 1$ and $D_O = K$. We set $W_I = 32$, $D_I = 128$, $K = 128$, $F = 3$, $P = 1$, and $S = 1$. Therefore, $W_O = 32$ and $D_O = 128$. In the baseline implementation, each cluster computes one depth slice (aka "channel") of the output volume (of dimensions $W_O \times W_O$) at a time. As the entire input volume does not fit into the local memory of a cluster, the cluster loads a stack of depth slices of the input volume at a time. Thus, each cluster needs to load the entire input volume once per output depth slice. As the first result column in Table 3.4 shows, this implies a very low operational intensity and entails that performance is bound by the HBM memory bandwidth. One strategy to alleviate this is to let each cluster compute a stack of depth slices of the output volume. As the input depth slices can be reused for multiple output depth slices, this reduces the amount of data transferred per computation. For a stack of 8 output depth slices (second column), the operational intensity is sufficiently high that the performance becomes compute-bound. To save even more off-chip bandwidth without sacrificing performance, the hierarchical network can be used to form a processing pipeline where clusters obtain their input depth slice from another cluster instead of off-chip memory. The third column shows that when all 16 clusters within one L2 quadrant form such a pipeline, the off-chip memory traffic can be massively reduced while performance is maintained. Traffic is also reduced on the L2 and L3 networks because data, once it is in the local memory of a cluster, is mainly transferred through the L1 networks.

A fully-connected NN layer transforms an input volume with dimensions $W_I \times W_I \times D_I$ to an output volume with dimensions $1 \times 1 \times D_O$. As any fully-connected layer can be represented as convolutional layer, we stick to the introduced notation and operations and set $W_I = 32$, $D_I = 128$, $K = 128$, $F = 32$, $P = 0$, and $S = 1$. Therefore, $W_O = 1$ and $D_O = 128$. As each filter parameter is used only once per input-output volume pair, fully-connected layers are usually implemented by transforming a *batch* of B input volumes to a batch of B output volumes; we use $B = 32$. Our implementation parallelizes the input depth slices over the clusters. Before the parallel section, each cluster allocates a private output volume and initializes it to zero. In the parallel section, the cluster first loads the entire batch of one depth slice of the input volume and then loops over the output depth slices. Within that loop, the cluster loads the filter parameters for the current pair of input and output depth slices and then enters an inner loop over the batch. Within the inner loop, the cluster multiplies the input depth slice of a batch element with the loaded filter parameters element-wise and then accumulates all products to a single value, which it adds to the output element for the current output depth slice and batch element. After the parallel section, the private output volumes of all clusters are reduced by summation to a single output volume, which contains the contributions of all input depth slices. As the last column in Table 3.4 shows, this implementation is compute-bound. There is no communication between the clusters in the parallel region because there is no data common to any two clusters. As such, the hierarchy of the network is not exploited, but the high bandwidth between HBM and any cluster is: it allows reaching compute-boundedness already for a batch size of 32. Larger batch sizes further reduce the off-chip bandwidth.

3.4 Related Work

NoC topologies, routing algorithms, flow control schemes, and router architectures have been subject to a vast amount of research (see [PD10; FB10; Jer+17; KC18; DT03; BD06] for detailed reviews). Important conclusions from this research are that the optimal on-chip network topology highly depends on the target application and computer architecture, and that routing strategies and flow control schemes are intertwined with the communication protocol, which all connected modules need to adhere to. Thus, we do not try to innovate in this field. Rather, the modules in our platform allow to build an on-chip network with arbitrary topology that adheres to a state-of-the-art, industry-standard protocol, following the paradigm put forward by application-specific NoC research efforts (see [CF16] for an up-to-date survey). Additionally, our elementary modules allow to design custom network modules, including custom endpoints such as caches and memory controllers, without having to deal with all protocol intricacies. To the best of our knowledge, our work is the first on-chip communication platform that offers elementary modules smaller than crossbars or switches.

Non-coherent on-chip communication is central for heterogeneous, accelerator-rich SoCs [GMC18b]. Protocols similar to AMBA AXI5 [AXI-F.b], which our platform directly supports, are IBM's CoreConnect [IBM07], Silicore's Wishbone [Sil10], Accellera's Open Core Protocol (OCP) [Acc13], and SiFive's TileLink Uncached Heavyweight (TL-UH) [SiF19]. They all, like AXI, are royalty-free standards. CoreConnect, Wishbone, and OCP provide a subset of the features of AXI5, and while they had been used in the past, they are nowadays not nearly as widely used as AXI. TL-UH, like AXI5, supports burst transactions, multiple outstanding transactions, and transaction reordering and uses valid-ready flow control. TL-UH has stricter forward progress requirements than AXI5, which our modules could also fulfill. While the specifications define interfaces and protocols for on-chip communication, they do not describe the architecture of network modules implementing them; that is an important contribution of our work. The OpenSoC Fabric [Fat+16] is an open-source implementation of a custom non-coherent protocol, with an interface to AXI-Lite in development. AXI-Lite does not support bursts or transaction reordering and is therefore not suited for high-performance communication. The ESP project [GMC18a] provides an open-source implementation of a 2D-mesh NoC with a custom protocol. Similarly, the non-coherent BaseJump Manycore Accelerator Network, which has first been used in the Celerity chip [Dav+18], adheres to a custom protocol and is designed for 2D-mesh networks. In contrast, our platform is topology-agnostic and adheres to an industry-standard protocol. An overview of on- and off-chip interconnects for NN accelerators is presented in [Nab+20]. They highlight the need for non-mesh topologies in NN accelerators, to which we contribute with our case study and topology-independent platform.

Commercial IP offerings for AXI exist from multiple vendors, and we compare with them in Table 3.5. Our work is the only one (1) whose architecture is fully disclosed in literature, (2) whose RTL code is open-source and modifiable, enabling, e.g., the exploration of arbitration algorithms with certain guarantees inside standard-compliant networks, and (3) whose area and timing (AT) characteristics may be disclosed not only on FPGAs. Despite its research origin, the implementation behind this work powers on-chip communication of an increasing number of application-specific integrated circuits (ASICs) (e.g., [ZSB20; Zar+19]). From a technical perspective, our work is the only one that offers

Product	Public Prod. Spec.	Open Architecture	Disclosable AT Results	RTL Open-Source	RTL Modifiable	Usable on FPGA	Usable on ASIC	Elems. for Custom IPs	Data Width [bit]	Max. Concur. Txns.	Data Width Conv.	ID Width Conv.	DMA Engine	OCM Controller
Arm [23] NIC-400	✓	✗	✗	✗	?	✓	✓	✗	32–256	1–127	✓	✓†	$	$
Arteris [24] FlexNoc	✗	✗	✗	✗	?	?	✓	✗	?	?	?	?	?	?
Synopsys [25] DW_axi	✗	✗	✗	✗	$	✓	✓	✗	$\leq$ 512	?	?	?	$	?
Xilinx [26] AXI LogiCore	✓	✗	✓	✗	?	✓	✗	✗	32–1024	1–32	✓	✗	✓	✓
This Work	✓	✓	✓	✓	✓	✓	✓	✓	8–1024*	1–128¶	✓	✓	✓	✓

? = not disclosed publicly. $ = licensed separately. † Auto-inferred inside crossbar, cannot be customized or instantiated standalone. * Limited by AXI standard, larger data widths theoretically possible. ¶ Range evaluated in this work, larger values theoretically possible.

Table 3.5: Commercial IP offerings for AXI compared with this chapter.

elementary modules (i.e., *network* (de)multiplexers) that can be used to build custom AXI-compliant IP modules without having to deal with all intricacies of the protocol, instead of only crossbars as finest-granularity modules. Additionally, our work supports all standard-defined data widths, supports the highest number of concurrent transactions, and comes with communication modules such as ID width converters, a DMA engine, on-chip memory controllers, and a last-level cache, which are licensed separately or not available at all from commercial vendors.

Cache-coherent on-chip communication protocols currently in use include Intel's UltraPath Interconnect [Mul17], AMD's scalable data fabric [Bur+19], IBM's Power9 on-chip interconnect [Sad+17], AMBA AXI Coherency Extensions (ACE) [AXI-F.b], AMBA5 CHI [CHI-D], and TileLink Cached (TL-C) [SiF19]. ACE and TL-C are extensions of AXI and TL-UH, respectively. As such, our platform could be extended for coherent communication by adding channels, transactions, and properties defined by these specifications. Under the ordering rules **(O1–3)**, coherence transactions cannot use the existing channels of regular transactions: In coherent communication, a regular transaction can entail coherence transactions, which must complete before the regular transaction can complete. Guaranteeing this would require different ordering rules. The other protocols are standalone specifications with very different properties. For instance, we refer to [Cav+20] for an open-source bridge for connecting to CHI from AXI. With such a bridge, our platform can connect to a coherent system interconnect if needed, possibly extending to multiple chips. Coherency in on-chip networks has been studied extensively in research, e.g., [EPS06; EPL08; APJ09]. A prominent system example is SCORPIO [Day+14], where a coherent mesh NoC interconnects 36 homogeneous cores on a die. Their work focuses on the NoC and router architecture for a coherent homogeneous multi-core, while we design an end-to-end non-coherent on-chip communication platform suitable for heterogeneous many-cores. Generators for cache-coherent on-chip networks have been presented in multiple works: Open2C [But+18] contains a library of modules for coherent networks written in Chisel. Like us, they present an LLC, which is separated from a coherence directory. In their 512 KiB L2 cache, the area overhead of control logic and buffers is 38 %, whereas an identical parametrization of our LLC has only 3 % overhead. The Rocket chip generator [Asa+16] constructs SoCs written in Chisel, and the coherent NoC adheres to TL-C. OpenPiton [Bal+16] generates tile-based manycore processors with a 2D mesh coherent NoC. One tile has an area of 1.17 mm^2 when targeting IBM's 32 nm SOI process at 1 GHz. Of the tile area, 22.3 % are occupied by 32 KiB of distributed L2 cache and directory controller and 2.7 % by the 5×5 NoC router. Accounting for one full technology node difference, the equivalent area in GF22FDX would be ca. 660 kGE and 80 kGE for 32 KiB L2 cache and the NoC router, respectively. The control logic of their L2 cache is ca. 3.3 times larger than that of our LLC, which could be due to the cache directory. Their 5×5 NoC router (without any virtual channels) is about the same size as a 5×5 configuration of our crosspoint (with up to 16 reorderable IDs). Open2C and OpenPiton implement a custom protocol, which complicates connectivity with third-party modules, whereas we adhere to an industry-dominant protocol. The modules in our work are implemented in synthesizable SystemVerilog, so they could be integrated into a higher-level generator as well.

3.5 Summary

This first fully open-source[3] platform for high-performance on-chip communication enables the construction of heterogeneous many-core and accelerator-rich SoCs independent of proprietary on-chip networks IPs. The platform advances the technical state of the art through two main contributions: First, network (de)multiplexers as elementary components make the design and verification of custom network modules substantially easier. Second, an end-to-end palette of modules from a DMA engine to on-chip memory controllers, including data and ID width converters, as well as the widest range of data widths and concurrent transactions enables new designs. For example, we designed and implemented a state-of-the art 1024-core MLT accelerator in a modern 22 nm technology, where our communication fabric provides 32 TB/s cross-sectional bandwidth at only 24 ns round-trip latency between any two cores. Future work enabled by our platform includes design space exploration and optimization of on-chip networks, networks designed for stringent application constraints (e.g., arbitration guarantees for

[3]SystemVerilog source code available under a permissive open-source license at `https://github.com/pulp-platform/axi`.

real-time execution), and co-integrated cache-coherent and non-coherent networks.

Chapter 4

Modular and Scalable Support for Atomic Operations in a Shared Memory Multiprocessor

Atomic memory operations (AMOs) are ubiquitous in modern concurrent algorithms. Many of them, such as compare-and-swap, fetch-and-add, and LR/SC, can be used to implement lock- and wait-free algorithms and data structures with strong progress guarantees [HS11]. Theoretically, lock-free algorithms allow an arbitrary number of threads to share a resource without the need for serial execution on a lock. This is paramount for scaling algorithms to a high number of threads, because even very short intervals during which threads are serialized drastically limit the potential speedup (Amdahl's Law) [HM08].

Even though most ISAs today define AMOs (e.g., x86 [Int19b], ARMv8 [Arm19b], RISC-V [Wat+11]), their scalable implementation is not a solved problem: First, implementation in commercial processors is a well-guarded secret, thus there is a knowledge gap on the challenges and trade-offs of implementing AMOs. Second, the subsystem for executing AMOs is presumably tightly coupled to the processor architecture and the memory hierarchy. Thus, techniques developed for one multiprocessor architecture do not readily apply to other architectures. Finally, AMOs on modern multiprocessors have been shown to scale poorly to large numbers of threads [SBH15; ELF11].

In this chapter, we fill the knowledge gap and present an open-source hardware module to implement AMOs at any level in the memory hierarchy. The proposed architecture decouples the execution of AMOs and conditional-store-based primivites from locking shared resources as much as possible. This allows our solution to scale the throughput of AMOs linearly until the target memory saturates.

Our module is designed for modern on-chip communication protocols (OCCPs) and ISAs in general, but for concreteness we describe an implementation compliant with the open, modular RISC-V ISA [Wat+11] (§ 4.1.1) and the industry-standard AXI OCCP [Arm17a] (§ 4.1.2). We describe how our module integrates into a memory hierarchy in § 4.2, give an overview of its design in § 4.2.1, describe the microarchitecture of its two stages in § 4.2.2 and 4.2.3, and discuss its liveness guarantees in § 4.2.4. We evaluate our system on a cycle-accurate FPGA prototype (§ 4.3.1), where 32 cores share a second-level scratchpad memory, and find (§ 4.3.3) that:

1. The throughput of AMOs scales linearly with the number of cores until the on-chip memory is saturated with and without contention;

2. The latency of an AMO is only 25 % higher than a regular load from that memory and under contention increases linearly with 10 cycles per core;

3. The throughput of important concurrent algorithms scales linearly with the number of cores until the memory bandwidth is saturated.

We furthermore synthesize our design for a 22 nm FDSOI technology for a variable number of cores in the system and find that its area increases linearly at only 0.5 kGE per core and its longest path scales logarithmically with the number of cores (§ 4.3.4). We compare to related work in § 4.4 and conclude in § 4.5.

4.1 Background

We briefly describe RISC-V's semantics for atomics and its memory consistency model (§ 4.1.1) as well as modern on-chip communication at the example of AXI (§ 4.1.2).

4.1.1 RISC-V ISA

RISC-V [Wat+11] is an open ISA and its flexibility and modularity make it ideally suited for this chapter. We follow the RISC-V terminology and call a processor component *core* if it contains an independent instruction fetch unit. One core might support multiple hardware threads (*harts*) through multithreading.

RISC-V's 'A' extension specifies two different types of atomic instructions: the LR/SC pair and AMOs. A load-reserved (LR) loads a word and simultaneously places a reservation for the hart on the read memory location. This reservation does not prevent other harts from reserving the same location or from reading or writing to the same location, but any modification of a memory location clears all reservations to that location. The store-conditional (SC) instruction stores a word at a memory location if the hart has a valid reservation and returns a value indicating whether the store succeeded. Together, the LR/SC pair can be used to implement atomic read-modify-write (RMW) operations.

Atomic memory operations (AMOs) implement a fixed set of atomic RMW operations, which match those of the C11/C++11 atomic operations library, facilitating its fast implementation. The operations are SWAP, ADD, bitwise AND, OR, and XOR, and signed and unsigned MIN and MAX on 32- and 64-bit words (the latter only on 64-bit processors).

RISC-V's memory model is built around release consistency [Gha+90]. Each hart executes its instructions so that they appear in program order as seen from the executing hart. Memory instructions from other harts, however, may be observed in a different order. This so-called RISC-V Weak Memory Order (RVWMO) gives computer architects the flexibility to design scalable and high-performance systems but burdens software developers with inserting explicit memory instructions where required – although RVWMO and AMOs were designed to implement the C11/C++11 memory model efficiently. RISC-V optionally defines a stronger total store ordering (TSO) memory model. Our work can accommodate both RVWMO and RVTSO.

4.1.2 Modern On-Chip Communication and AXI

Modern on-chip communication is centered around the premise of high-bandwidth point-to-point data transfers. To fulfill this premise despite increasing point-to-point latencies, three central traits of modern on-chip communication protocols are: *burst-based transactions*, *multiple outstanding transactions*, and *transaction reordering*. Our design targets these central traits in general, so the concepts we present potentially apply to a wide range of modern on-chip protocols. More tangibly, we adhere to the latest revision (5) of the AMBA Advanced eXtensible Interface (AXI) [Arm17a]. AXI is one of the industry-dominant OCCPs and the only OCCP with an open specification and a widespread adoption in current real-life systems designed by many different companies. Other modern major commercial OCCPs with similar properties include Intel's Ultra Path Interconnect [Mul17], AMD's scalable data fabric [Bur+19], and IBM's Power9 on-chip interconnect [Sad+17].

AXI separates communication into two directions (read and write), into channels for commands, data, and responses, and into transfer items called 'beats'. A transaction starts with a command followed by one or multiple data beats and ends with a single response (on a write) or the last of multiple responses (on a multi-beat read). Each transaction is initiated by a master, targets an addressed slave, and can involve multiple interconnecting components.

Exclusive accesses in AXI are very close to the semantics of LR/SC in RISC-V. The basic mechanism is that a master issues an exclusive read (LR in RISC-V) to an address and some (unrestricted) time later an exclusive write (SC) to the same address. The exclusive write then succeeds if no other master has written to that address since the exclusive read and fails otherwise. Only successful exclusive writes modify the memory.

Atomic transactions, which are write transactions with an added atomic opcode, were added in AXI5. There are four types of atomic transactions: store, load, swap, and compare. An atomic swap unconditionally replaces the memory value at an address with the provided data value and returns the original value; an atomic compare replaces the value in memory only if it matches a second provided data value. Atomic loads and stores unconditionally apply one of eight operations (add, clear, exclusive or, set, and signed and unsigned minimum and maximum) on the memory and a provided value. Atomic loads return the original memory value; atomic stores return a response without data. Although specified independently, the atomic transaction operations of AXI5 are a superset of the AMOs of RISC-V: the atomic ADD, AND, OR, XOR, and signed and unsigned MAX, and MIN instructions can be mapped to atomic loads (or atomic stores if the destination register is `x0`) and the SWAP instruction to an atomic swap.

4.2 Design and Architecture

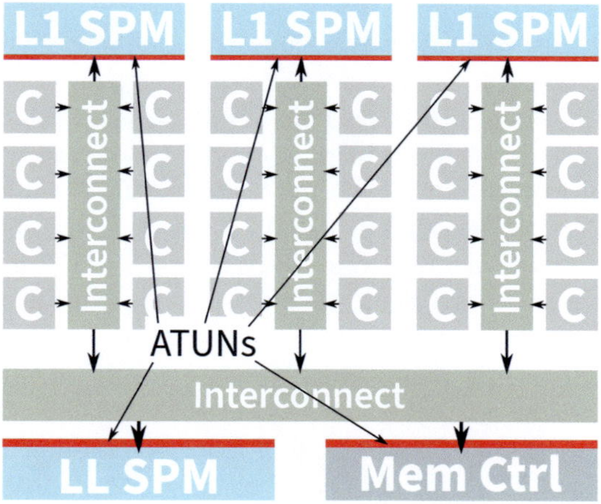

Figure 4.1: Implementing AMOs with the ATUN (red horizontal bar) in a memory hierarchy based on scratchpad memories (SPMs).

We describe the design and microarchitecture of our ATomic UNit (ATUN) hardware module. Fig. 4.1 shows how the ATUN can be placed in front of any memory that has an AXI-like OCCP interface, giving system designers many options on where in the memory hierarchy to resolve AMOs and LR/SCs.

4.2.1 Design Overview

Our ATUN is composed of two stages: the AMO stage, which serves the OCCP slave interface, and the LR/SC stage, which controls the master interface (left ports of Fig. 4.2). The AMO stage (§ 4.2.3) resolves AMOs in its arithmetic logic unit (ALU) and uses the atomicity guarantee provided by the subsequent LR/SC stage to guarantee the single-copy atomicity of each AMO. The LR/SC stage (§ 4.2.2) guarantees the atomicity of an LR/SC pair for any reorderable downstream memory interface.

4.2.2 LR/SC Stage

The LR/SC stage guarantees the single-copy atomicity of LR/SC pairs as long as it "owns" the entire downstream memory – i.e., no transactions can reach the downstream memory without being observed by the LR/SC stage – but the downstream memory (e.g., an off-chip memory controller) is free to reorder transactions as defined by the OCCP's memory semantics.

The LR/SC stage will always only emit transactions that are non-exclusive. Downstream modules unconditionally execute all writes and have full freedom of reordering transactions (as allowed by the OCCP specification), which is essential for memory controllers and other off-chip links to achieve high bandwidth. The LR/SC stage considers these reordering options and fails an exclusive store if a contending write could be reordered before the exclusive store. A central feature of the LR/SC stage is that it does *not lock* the write channel during exclusive accesses.

The LR/SC stage, shown in Fig. 4.2, is composed of a reservation table and control FSMs for the OCCP channels, which interact through command queues. Read and write transactions are fully independent. The algorithm to process them is essentially as follows: Every read request is forwarded in the same clock cycle and, if it is exclusive and no write to the same address is in-flight downstream, places a reservation. If both a read and a write request to the same address region are at the upstream interface simultaneously, the write is stalled for one clock cycle to order the effect of reads and writes on reservations. To maintain the single-copy atomicity of LR/SC, a write is also stalled if an address-overlapping exclusive write is in-flight downstream. In

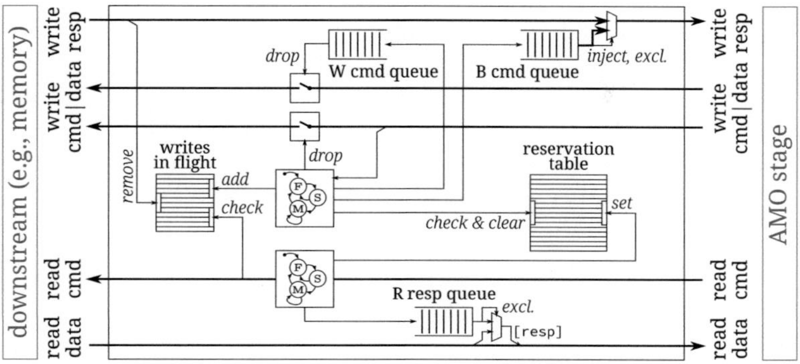

Figure 4.2: Microarchitecture of the LR/SC stage.

all other cases, a write request is forwarded in the same clock cycle. Exclusive write requests are forwarded if and only if the reservation table holds a reservation for the targeted memory range and the hart identified by the transaction ID. Each forwarded write request clears all reservations to overlapping address ranges. Our design supports any order of LR/SCs from the connected harts, and all harts can reserve any address in the attached memory at any time. When placed before a cache, the LR/SC stage relies on the existing coherence protocol to guarantee atomicity and snoops the coherence channel for invalidations, upon which it clears matching reservations. If there are caches before the ATUN, they, upon an AMO must invalidate the affected cache line and forward the AMO.

The reservation table is at the core of the LR/SC stage. Each hart is identified by a unique OCCP transaction ID, and since the reservations by different harts must be independent, the reservation table contains one entry per hart. While a cache-like structure with less entries than harts would be attractive to save area, doing so adds dependencies between reservations from different harts, which can lead to livelocks and deadlocks. The table is indexed by the ID and stores the start address and size of an exclusive access.

The reservation table has two interfaces: The first interface checks if a hart holds a reservation for an address and optionally clears all reservations that have a range that match the address. The second interface sets the entry for a hart to an address and size. In the worst case (overlapping reservations for all harts and a write to an address in the intersection), the entire table has to be read and written before a request can be granted. As latency is crucial for the LR/SC stage (discussed in § 4.3.3), the table is implemented as an array of flip-flops so that all entries can be compared and modified in parallel.

4.2.3 AMO Stage

The AMO stage executes AMOs by leveraging the atomicity guarantee of the LR/SC stage. In a nutshell, each AMO is translated into the following transactions: First, a read is issued and the returned data is fed to the internal ALU together with the operand. After the ALU has computed the result, a write is issued to store it back to the memory. Once the write response asserts that the operation is complete, both a write response and a read response, containing the previous memory value, are sent to the issuer of the AMO.

Downstream Transactions

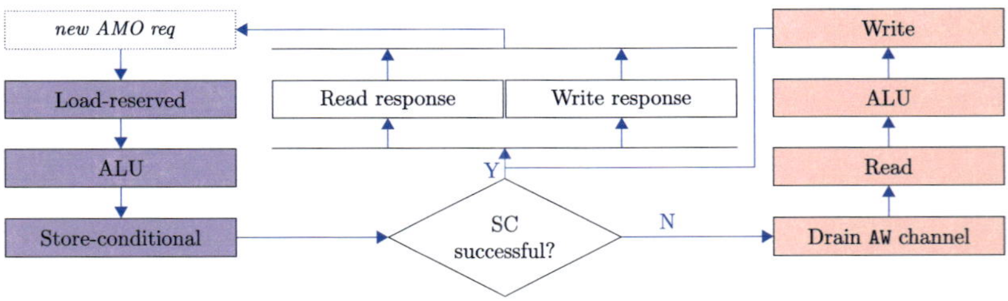

Figure 4.3: Flow diagram of AMO with *fast path* (blue, left) and *slow path* (red, right).

As the OCCP does not have to order read with respect to write transactions, a read issued by the AMO stage can overtake a write to the same address that passed the AMO stage before it issued the read. In this case, the write would violate the single-copy atomicity of an AMO. The LR/SC stage solves this problem for SCs, and instead of replicating the logic, the AMO stage uses the guaranteed atomicity of a successful SC to ensure the atomicity of an AMO. As illustrated in Fig. 4.3, as a first step on the *fast path*, an LR is issued to resolve an AMO. Upon receiving the data and calculating the result, an SC is used to write it back to the memory. If the SC was successful, the LR/SC stage guarantees the atomicity of the LR/SC pair and the AMO stage can send the responses for the AMO. This sequence is called the *fast path*, as it executes the AMO immediately with the assumption that it will succeed. However, an SC on the *fast path* might fail due to conflicting writes by non-AMOs downstream, but an AMO must never fail. Therefore, if the SC fails and does not update the memory, the AMO stage executes the *slow path*.

The *slow path* only occurs when a program updates a memory location both with regular writes and AMOs (which is not data race free). This special case is usually not worth optimizing for, and our implementation minimizes hardware spent on it: First, the AMO stage completely drains the downstream write channel to eliminate the possibility of a conflicting write transaction in flight. Second, the AMO stage stalls not only conflicting but all new write requests until the AMO has executed, because it does not keep track of the target addresses of writes in flight and there are no ordering guarantees on writes. Finally, the AMO stage uses a regular read followed by a regular write transaction to execute the AMO.

Microarchitecture

As shown in Fig. 4.4, the microarchitecture consists of controllers for the OCCP channels and an execution unit to compute AMOs. In the absence of AMOs, the AMO stage is transparent for incoming transactions. When handling an AMO, the AMO stage injects reads and writes between regular transactions with priority. This not only reduces the latency of single AMOs but also increases the throughput of AMOs, as this microarchitecture processes AMOs sequentially. This microarchitecture was designed for low latency and low area, but architectural extensions that feature multiple parallel execution units, pipeline the fetching and writeback of operands, and/or can fuse AMOs to the same address are possible.

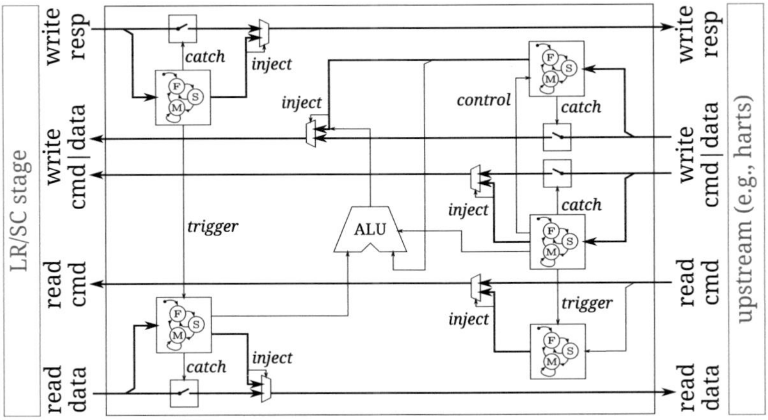

Figure 4.4: Microarchitecture of the AMO stage.

	starvation-free	livelock-free	deadlock-free
AMO	✓ (D)	✓ (B)	✓ (A)
LR/SC	✓ (E)	✓ (C)	✓ (A)

Table 4.1: Liveness properties guaranteed by our ATUN.

4.2.4 Liveness Guarantees

Three liveness properties are essential to guarantee progress in multi-master communication: freedom from starvation, from livelocks, and from deadlocks. OCCPs stipulate rules to make any two transactions free from livelocks and deadlocks, and arbitrators are commonly required to be starvation free. Under these assumptions on downstream and upstream, our ATUN guarantees the liveness properties listed in Table 4.1 as follows: (A) Both AMOs and LR/SCs are free from deadlocks because each AMO, LR, and SC individually completes within a bounded number of cycles and, once completed, does not preclude any other instruction from progressing. (B) AMOs are entirely free from livelocks because they are executed in the order they enter the AMO stage, resulting in unconditional progress among all AMOs. (C) LR/SCs are free from livelocks because one among multiple contending LR/SC pairs always succeeds. (D) AMOs are starvation-free because they are unconditionally executed in the order they enter the AMO stage within a bounded number of cycles. (E) The execution of LR/SC pairs is free from starvation, but their success is only guaranteed to be starvation-free for disjoint addresses because any such pairs do not change the success of each another. While it is possible to write programs based on LR/SC that prevent one hart from ever succeeding an SC (or that deadlock or livelock the program), such liveness violations do not extend to other programs or even system components and thus do not impair the liveness guarantees of our ATUN.

4.3 Evaluation

We built a multicore architecture (§ 4.3.1) to evaluate multiple variants of four benchmarks representing a wide range of loads on our ATUN, and we characterize throughput and contention (§ 4.3.3) in cycle-accurate execution on an FPGA. Finally, we characterize the hardware complexity in a 22 nm technology (§ 4.3.4).

4.3.1 Evaluated Architecture

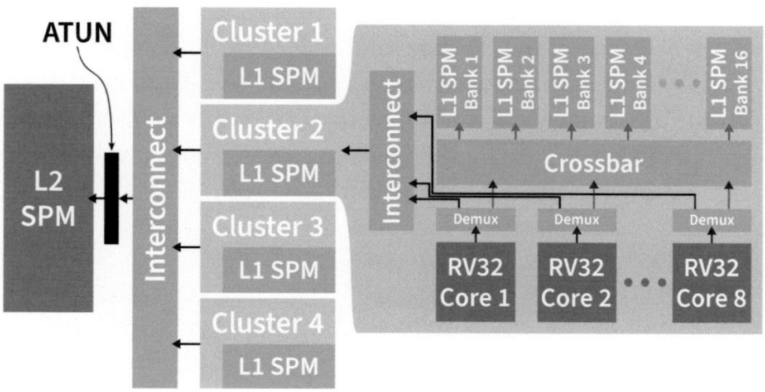

Figure 4.5: Architecture of the evaluated system, where one instance of our ATUN is in front of the L2 SPM shared by four clusters each composed of eight RISC-V cores.

Fig. 4.5 shows an overview of the multicore architecture that we assembled to evaluate our ATUN. 32 cores organized in 4 clusters share a L2 SPM through a fully-connected-crossbar interconnect. In front of the L2, one ATUN handles atomic and exclusive memory accesses. We focused our evaluation on this case to show the benefits and limits of our proposed approach where many harts share one ATUN. All cores within a cluster share a multi-banked L1 SPM, which is used by the benchmarks to store core-private data. Each core implements one RISC-V in-order hart.

We implemented the evaluation architecture on a field-programmable gate array (FPGA) to be able to measure throughput and contention in benchmarks cycle-accurately. For measurements inside the memory hierarchy, we inserted hardware performance monitors that do not interfere with execution into our system. We measured the number of cycles on the cycle-accurate FPGA implementation and scaled throughput and latency numbers (§ 4.3.3) to the frequency achieved by the ASIC implementation (§ 4.3.4).

4.3.2 Terminology: Atomic Locality

We call a set A of atomic variables *local* to a set of harts H during a time interval T if there exists no hart outside H that accesses any variable in A during T. Let H consist of all harts executing a workload, then during some interval T, the *atomic locality* of that workload is $|H|/|A|$. A high atomic locality is neither "good" nor "bad": For highest performance, the memory to which the ATUN is connected should be able to simultaneously hold all variables in A, implying moderate values of atomic locality are "good". If the atomic locality is too high, however, it becomes "bad" as the probability of conflicts in the shared ATUN increases.

4.3.3 Throughput and Contention

We selected four different benchmarks to evaluate our ATUN under a wide range of loads. In all benchmarks, contention is maximized by programming all harts to execute the specified atomic operations without interruption. The execution time of the slowest hart is used as the total system execution time.

Synthetic Maximum Contention

For maximum contention, a variable number of harts execute a single (write or AMO) or two (LR and SC) memory operations to the L2 memory without interruption in a loop. All writes and AMOs target the same memory location. All LR/SCs go to different memory locations so that every SC is successful and leads to a write. We use this to find the upper bound on the throughput and the lower bound on the latency as a function of harts under maximum contention.

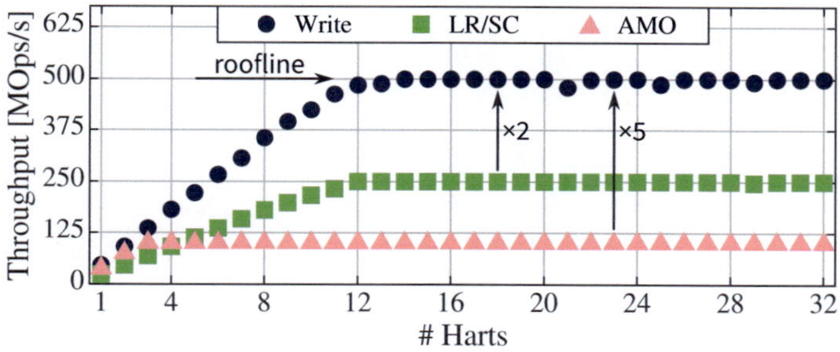

Figure 4.6: Throughput of writes, LR/SC pairs, and AMOs as a function of harts under maximum contention.

The throughput of writes, LR/SC pairs, and AMOs under maximum contention on the evaluated architecture is shown in Fig. 4.6. Even though the memory is clocked at 1 GHz, the memory controller can only accept a read or write every second clock cycle, leading to a peak throughput of 500 MOp/s. The write curve forms the roofline for the evaluated architecture. Both writes and LR/SC pairs scale linearly up to 12 harts, and LR/SC pairs achieve exactly half the throughput of writes because one LR/SC pair is composed of two memory operations. The current implementation of our ATUN requires 10 cycles to resolve one AMO if the memory controller can accept a read or write only every second cycle: 2 cycles forth and 2 cycles back for the read, one for the computation of the AMO, four cycles for the write, and one to accept the next AMO. Therefore, it can process one AMO every tenth cycle, which is five times lower than writes only. Nonetheless, the evaluated architecture can sustain a throughput that is half the roofline for LR/SCs, which is optimal when all SCs are successful (as in this benchmark), and a fifth of the roofline for AMOs even under maximum contention, which is a synthetic worst-case scenario.

Lock-Free Memory Allocator

We implemented the lock-free buddy allocator proposed in [Mar+18]. Memory is allocated in chunks of discrete sizes and managed by a binary tree, whose nodes are atomic variables. We implemented the algorithm once with AMOs and once with some AMOs replaced by LR/SC-based atomic read-modify-writes to maximize throughput. This algorithm requires modifying several different shared variables atomically, so it represents algorithms with low atomic locality.

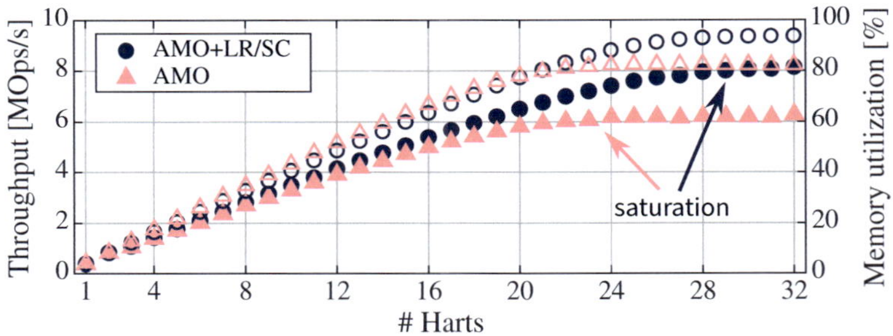

Figure 4.7: Throughput (left scale, filled markers) and utilization of the memory interface (right scale, empty markers) as a function of harts for two variants of the lock-free memory allocator, which has low atomic locality.

The throughput of the lock-free allocator is shown in Fig. 4.7. The AMOs-only variant scales linearly up to ca. 20 cores. Beyond, the ATUN saturates as it can handle one AMO every 10 cycles. The second variant combines AMOs and LR/SC. This leads to a balanced utilization of the ATUN *and* the memory, saturating both for the maximum number of harts.

Lock- or Wait-Free Concurrent Queue

As a representative of algorithms with high atomic locality, we implemented the concurrent queue algorithm proposed in [YM16]. Only two atomic variables, the head index and tail index, arbitrate the access to the shared queue. We implemented two variants of the algorithm: in one we use AMOs to atomically modify the head and tail index, in the other we use LR/SC pairs for that purpose. As the variant with AMOs is wait-free whereas the variant with LR/SCs is only lock-free, we use this to compare the throughput of the two operation types and the cost of the higher progress guarantee.

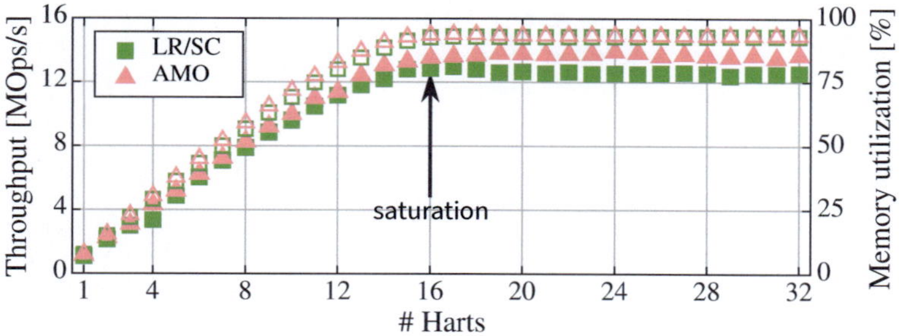

Figure 4.8: Throughput (left scale, filled markers) and utilization of memory interface (right scale, empty markers) as a function of harts for two variants of the concurrent queue, which has high atomic locality.

The throughput of both variants is shown in Fig. 4.8. Remarkably, the throughput of the variant with the stronger progress guarantee (AMO) is consistently higher. The reason is that this concurrent queue has a very high atomic locality with only two atomic variables. In such cases, the success rate of LR/SC pairs quickly degrades and limits the throughput of the LR/SC variant whereas AMOs always succeed. The algorithm reads from L2 memory not only with AMOs, and the combined throughput of reads limits the AMO variant.

Parallel Histogram

The histogram benchmark is representative for algorithms with data-dependent atomic access patterns. The shared target histogram is allocated with B bins in the L2 memory. Each hart reads values from an array in the L1 memory of its cluster and atomically increments the bin to which the value belongs. We implemented two variants, one that uses LR/SC and one that uses an AMO ADD for atomically modifying the shared variable, to be able to compare the effectiveness of LR/SC to AMOs in this scenario. In order to focus on L2 contention, we deliberately do *not* accumulate in the shared L1 and then update the L2 in batches.

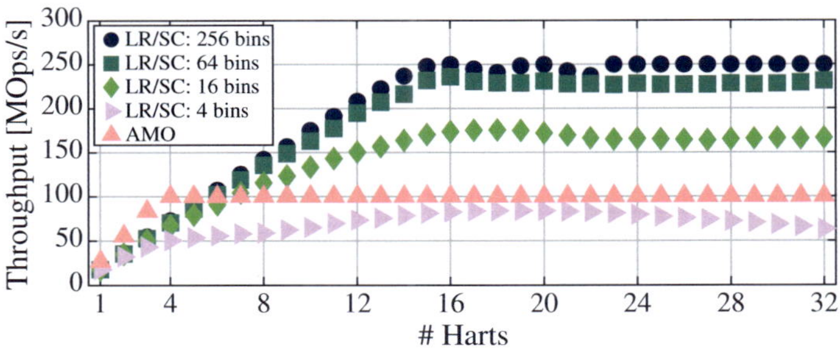

Figure 4.9: Throughput of the histogram benchmark as a function of harts for different atomic localities.

The throughput of AMOs and LR/SCs in the histogram benchmark is shown in Fig. 4.9. As the current implementation of the AMO stage processes all AMOs in series, the locality does not influence the throughput. For LR/SCs, however, it is of major importance, because their success depends on the number of shared variables accessed. This benchmark shows the locality of AMOs required for LR/SC to outperform AMOs under maximum contention. With the likelihood of collisions decreasing when more bins are used, the LR/SC version reaches the maximum possible throughput on the evaluated system at 250 MOp/s.

Summary

In most cases, the throughput of AMOs through the ATUN scales linearly until the memory at which the ATUN executes AMOs saturates. When the throughput of AMOs saturates before the memory bandwidth is reached, this is due to the AMO stage processing one AMO every 10 cycles, and further microarchitectural improvements could alleviate this bottleneck. Whether to use AMOs or LR/SCs to implement an atomic operation depends mainly on the atomic locality.

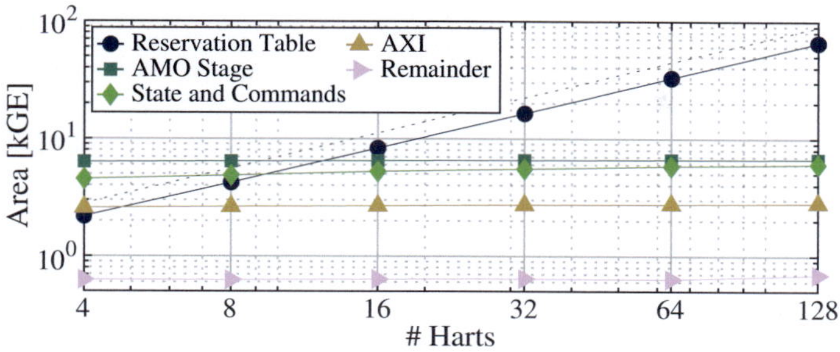

Figure 4.10: Hardware complexity of our ATUN as a function of the maximum number of concurrent harts. The dashed line is $y = 0.7x$ to show the linear growth of the reservation table at 0.5 kGE per hart.

4.3.4 22nm FDSOI Hardware Complexity

We synthesized our ATUN for a target frequency of 1 GHz on GlobalFoundries' 22 nm FDSOI. Fig. 4.10 shows the hardware complexity in number of gates of one instance of our ATUN as a function of the number of harts for which it can track reservations. Overall, the complexity increases linearly with the number harts—both axes are log scale—at only 0.5 kGE (100 µm^2 in 22 nm) per hart. The complexity is dominated by the reservation table, which causes the linear asymptote. The OCCP interface grows logarithmically with the number of harts (to encode additional hart IDs), and the state and command queues as well as the entire AMO stage remain constant.

The longest path through our ATUN (ca. 1050 ps) involves the unpipelined ALU in the AMO stage, and it does not increase with the number of harts. Few paths grow logarithmically with the number of harts, e.g., at the interfaces of the reservation table, but they do not become critical for the evaluated number of harts. In summary, the hardware implementation of our ATUN is ideally suited to scale to a large number of harts at a very low hardware cost per hart.

4.4 Related Work

To the best of our knowledge, our work is the first to introduce the concept of *decoupling* reservations for exclusive memory accesses and execution of AMOs. As a result, for the first time memory hierarchy designers can decide at which memory level AMOs are resolved. Additionally, our work is the first to describe the microarchitecture of a hardware module that is designed to be shared by multiple harts to resolve atomic memory accesses and evaluate its performance scalability and hardware complexity. As AMO resolution is deeply embedded into the memory subsystem in related work, a quantitative comparison is very complex and we focus on a qualitative comparison.

Executing atomic fetch-and-ϕ operations close to the memory has already been proposed in early multiprocessors [Got+83], and our work extends this concept to LR/SCs and modern on-chip communication with transaction reordering. Combining networks have been proposed to scale fetch-and-ϕ to many parallel requestors [Got+83; Tze92; LS94], and while our current ATUN cannot directly be integrated into network nodes, extending our work in this direction potentially allows to scale the throughput beyond the bandwidth of the target memory.

The x86 ISA [Int19a; Int19b] specifies AMOs that are different from those of RISC-V, and these operations seem to be resolved at the L1 cache of each core [SBH15]. However, there is no public information available on how write buffer, L1 cache, and pipeline stages interact to resolve AMOs. The poor performance of RMWs in x86 is partially caused by TSO requirements, and works such as [Raj+13] that relax those are orthogonal to our work. ARM [Arm19b] specifies similar AMOs as RISC-V and their AXI and ACE [Arm17a] protocols allow remote AMOs, but there is no public information available on the options for resolving atomic memory operations in an ARM-based multiprocessor.

Two instances of open-source, many-core capable multiprocessors are OpenPiton [Bal+16] and Rocket chip [Asa+16]. In OpenPiton, each core has a private L1 and L1.5 cache, and all cores share a common (distributed) L2, at which AMOs are executed. In contrast to the ATUN, AMO execution is embedded into the cache and does not include uncached or off-chip accesses. The multicore Rocket chip [Asa+16] uses a coherent TileLink-C interconnect [SiF18] similar to ACE. AMOs to the peripheral space are forwarded on the bus while atomics on main memory are resolved in the processor's L1 data cache by using the cache coherency protocol to obtain a unique copy of a cache line. This seems similar to the close coupling of resolving atomics between core and L1 on x86 architectures. In contrast, our ATUN could resolve AMOs at the last level cache.

GPGPUs are another important class of parallel processors, known in particular for scaling performance to a large number of threads. AMOs on memory shared by more than one SIMT unit, however, were long avoided by programmers of GPGPUs because their lock-based implementation [Gom+13] used to destroy just that performance scaling both on AMD [ELF11] and NVIDIA [Adi14] GPGPUs, and researchers proposed architectural changes to improve this [FL13]. Indeed, the latest GPGPUs generations significantly improved the performance scaling of atomic instructions by adding microarchitectural support for executing them on shared memory [De +19; Jia+18]. Nonetheless, atomic updates and synchronization remain a limiting factor in many HPC applications and the demand for faster atomics persists [HTE17].

4.5 Summary

We propose the concept of modular ATomic UNits (ATUNs), which decouple the execution of AMOs and conditional-store-based primitives from locking shared resources in the common case. ATUNs can implement AMOs at different levels of the memory hierarchy in manycore processors. We designed and implemented an ATUN that supports RISC-V's AMOs and memory model on standard OCCP interfaces (AXI specifically). We demonstrated the performance of our ATUN on a cycle-accurate FPGA prototype with 32 cores. We evaluated the hardware complexity of our design in 22 nm FDSOI and find that its area scales linearly at only 0.5 kGE per hart and its combinatorial delay scales logarithmically. Our ATUN has been integrated into the application-class open-source Ariane [ZB19] RISC-V core, which successfully runs Linux, where atomics play a vital role.

We expect that our work lays the foundation for further research on modular microarchitectures and optimizations of ATUNs. As explained throughout the architecture and evaluation sections, no single ATUN microarchitecture can perfectly match the wide range of multiprocessor architectures and domain-specific concurrent workloads. To foster future work on this topic, we release our ATUN implementation, which is written in industry-standard SystemVerilog, under a permissive open-source license at `https://github.com/pulp-platform/axi_riscv_atomics`.

Chapter 5

Scalable and Efficient Virtual Memory Sharing with TLB Prefetching and MMU-Aware DMA Engine

In HeSoCs, a general-purpose multicore CPU, the *host*, is co-integrated with PMCAs on a single die. This design holds the promise of combining the versatility of the host CPU with the energy efficiency and computing performance of the highly parallel accelerators.

One of the major difficulties in programming HeSoCs is having to explicitly manage the multi-level, non-uniform memory system. On the host, coherent caches and virtual addresses make the memory hierarchy completely transparent to the application programmer. On the PMCA, however, SPMs are often preferred to hardware-managed caches for the implementation of on-chip memory hierarchies. SPMs are physically addressed and data transfers to and from them are controlled by software, preferably using DMA transfers.

To alleviate this difficulty and to enable sharing of linked data structures, the Heterogeneous System Architecture Foundation [HSA12] pushed an architectural model where host and PMCAs communicate via coherent SVM. For coherency with the host data caches, most SoCs today offer accelerators access to coherent interconnects [Stu+15; Goo13]. For the translation of virtual addresses, there are two main approaches: In the all-hardware approach followed by many embedded SoC vendors, a full-fledged IOMMU translates addresses autonomously [PHB14; ARM16]. It is comprised of a TLB, parallel hardware page table walkers (PTWs), transaction and data buffers, and coherent page table caches. The alternative is a hybrid hardware-software design, which consists of a TLB controlled by software (e.g., by a kernel driver on the host [LAC14; VMB17] or directly by the accelerator [Vog+17]). We subsequently refer to the former class of IOMMUs as *conventional* and to the second class as *hybrid*. While conventional IOMMUs have the advantage of being transparent to the PMCAs and of offering the minimal latency for handling an isolated TLB miss, they have three significant drawbacks:

First, parallel, interleaved accesses by PMCAs to independent virtual addresses require parallel PTWs. While the number of parallel accesses is a time-variant run-time property of programs executed by the PMCAs, the number of PTWs is a fixed design parameter in conventional IOMMUs. To accommodate a wide range of parallel workloads, the number of parallel hardware PTWs must be overprovisioned, wasting hardware resources in most use cases.

Second, enabling DMA engines to access SVM in conventional IOMMUs requires a data buffer in the IOMMU to absorb write bursts to addresses that miss in the TLB. This buffer requires a significant amount of memory: it must have at least the size of the largest DMA burst, and is usually even larger because no more SVM accesses (by *any* master, not just the missing DMA engine) can be processed once (and as long as) that buffer is full.

Third, a conventional IOMMU manages its TLB purely reactively: new entries are set up only after a TLB miss. While some IOMMUs [Ves+16; ARM16] can speculate on future memory accesses based on past access patterns, doing so is very inaccurate for nonlinear, interleaved access patterns and negatively affects performance [Ves+16]. Misses can thus occur frequently, and high-performance conventional IOMMUs include coherent data caches for page table entries [ARM16] to reduce the latency of handling a TLB miss. If the TLB was managed by software threads in the PMCA instead (as in hybrid IOMMUs), it would be possible for those threads to set up TLB entries ahead of time based on run-time information inside the PMCA. Research on data caches [RS99; ABC03] has long shown that prefetchers that know the running program and its status can be far more accurate than those that can only see a stream of memory addresses.

The hybrid design, on the other hand, theoretically does not have these drawbacks. However, the state-of-the-art implementation [Vog+17] features only a single PTW thread, only supports DMA bursts that are guaranteed not to miss (e.g., by locking the corresponding TLB entries), and does not perform any prefetching.

In this chapter, we resolve these limitations. To the best of our knowledge, this chapter is the first to:

1. implement accurate, compiler-generated prefetching for a shared TLB (§ 5.3.1), which significantly reduces the rate of TLB misses,

2. offer a flexible number of parallel TLB miss handlers (§ 5.3.2), which keeps the miss handling latency constant for scalable parallel workloads, and

3. offer shared virtual memory accessible by DMA transfers without additional buffers (§ 5.3.3).

Compared to the state-of-the-art hybrid IOMMU [Vog+17], our contributions improve the PMCA performance for memory-intensive kernels by up to 4× and by up to 60 % for irregular and regular memory access patterns, respectively (§ 5.4.3). Compared to using data buffers to absorb bursts from DMA engines, our solution requires two orders of magnitude less memory (§ 5.4.4) and scales better, as it only stalls the missing DMA engine.

5.1 Related Work

The vast majority of commercial systems today features conventional IOMMUs [ARM16; Int15; Kor+14; Xil17] to completely abstract the SVM implementation from the PMCAs. While simple to use, that approach is limited in scalability (handling parallel misses and absorbing burst transfers) and efficiency (reactive TLB management).

Parallel TLB miss handling and page table walking The fixed number of shared hardware PTWs puts an upper bound on the scalability of conventional IOMMUs to parallel accelerators. A recent study [Ves+16] has shown that an integrated GPU with 8 parallel compute units (CUs) quickly saturates the miss handling capabilities of an IOMMU with 16 hardware PTWs, after which the GPU's performance becomes bounded by TLB miss handling latency. To avoid this, the current proposal for address translation on GPUs [PHW14; PHB14] is to add one MMU before the L1 cache in every CU. Each such CU MMU has its private TLB and either has its own PTW [PHB14] or shares a highly-threaded PTW [PHW14]. As this approach adds a significant amount of hardware, its parameters (e.g., TLB size, number of PTWs) must be carefully balanced *at hardware design time* to neither present a bottleneck for SVM-heavy applications nor reduce the compute-per-area ratio for applications that use SVM in a lighter way. The miss handling throughput of hybrid IOMMUs, where page table walks are performed by software threads, on the other hand, can be scaled *at run time* by scheduling PTWs when required. However, efficiently managing a TLB shared by many parallel PEs and notifying individual PE with low latency about handled misses is not trivial. For this reason, current hybrid SVM solutions [Vog+17; LAC14] only feature a single PTW thread. In this chapter, we show how to efficiently manage a shared TLB with multiple software PTW threads.

Handling bursts missing in the TLB The buffers in conventional IOMMUs that absorb write bursts missing in the TLB [ARM16] are another limiting factor for accelerators based on DMA transfers: When (and as long as) the limit on outstanding misses is reached, the IOMMU cannot translate any further transactions, even if they would hit. This creates backpressure from the IOMMU slave port to the connected master ports, stalling each master port on its next SVM access. Hybrid IOMMUs, on the other hand, signal the TLB miss back to the master and drop the transaction [VMB17]. This allows to handle misses on a shared TLB in a much more scalable way: instead of creating congestion on shared resources (e.g., buffers in the IOMMU, interconnect), the transaction that missed stays in the source memory, keeping shared resources clear for other accesses. To support this, the DMA engine must be able to keep track of bursts that missed and reissue them when the miss has been handled. In this chapter, we introduce a lightweight hardware extension that adds this feature for a standard DMA engine.

Reducing TLB misses through prefetching Reducing the number of TLB misses is another effective way to reduce the run time overhead of SVM, orthogonal to reducing the miss handling latency. There are two independent strategies to achieve this: The first is to increase the capacity of the TLB, for which both conventional and hybrid IOMMUs feature multi-level TLBs [ARM16; VMB17]. The second is to ensure the timelineness of TLB entries, e.g. through prefetching. Prefetching for shared TLBs is not yet well-understood: Some conventional IOMMUs feature a very simple prefetcher, which adds two subsequent pages to the TLB in case of a miss to the first [ARM16]. However, this prefetcher is deactivated by default because it harms performance in most cases [ARM16]. Prefetching is also supported by the PCIe Address Translation Services [PCI09], but a recent study [Ves+16] examined a benchmark with low locality, found that having a GPU prefetch eight contiguous pages degrades performance by up to 3×, and concluded that research on application-aware prefetching is required. In this chapter, we design and implement accurate prefetching for a shared TLB, which significantly reduces the rate of TLB misses. We focus on linked data structures (LDSes), which are the predominant source of scattered memory accesses in many programs. Our design is inspired by the following prefetchers for data caches.

Prefetchers that get information about the running program from software [RS99; LM99; KDS00; ABC03; GB06; EMP09; Lee+09; Son+09] are far more effective than heuristic hardware units [CJG02; Col+02] for LDSes: Heuristic hardware prefetchers for LDSes identify pointers as they are loaded from memory, prefetch their content before they are dereferenced, and store it in a separate pointer cache [Col+02] or in the data cache of the processor [CJG02]. All pointers identified by the heuristic hardware are prefetched recursively, which leads to a low prefetch accuracy, consequently polluting the cache and decreasing performance in many cases. To improve prefetch accuracy, the hardware prefetcher in hybrid hardware/software prefetchers [EMP09; ABC03] is controlled by software, e.g., from the main processor through special instructions to identify useful prefetches [EMP09] or by running a separate prefetching program [ABC03]. Prefetch code can be written manually by a developer [RS99; GB06] or generated automatically by a compiler [KDS00; LM99; Lee+09; Son+09], through static or dynamic profiling or both. Compilers can accurately identify pointers and prioritize their prefetching according to dependencies. Once prefetchers for LDSes are accurate, their effectiveness is limited by memory latency, as prefetch targets in LDSes depend on earlier pointer dereferences. As a dedicated hardware prefetcher co-located with the processor has the same memory latency as the processor itself, pure software prefetchers have been explored instead [GB06; LM99; Lee+09; Son+09]. Prefetches inserted inline with the actual program code [LM99], however, are limited to targets that are known when that line of code is executed. Otherwise, the actual program has to be stalled while the pointers leading to the prefetch target are followed. A promising alternative is to run an additional prefetching thread on the same multithreaded processor core [Luk01; GB06] or on another core in the same processor [Lee+09; Son+09]. Another important advantage of these separate prefetcher threads is that their throughput can be scaled to the demands of the application at compile-time or at run-time or both, especially for the high degree of parallelism offered by PMCAs. A key difficulty of software prefetches executed by another thread is the timeliness of the prefetches, which is why prefetching threads have primarily been explored for coarse-grained prefetching [Son+09; Lee+09].

While there are a number of works using heuristic hardware units for TLB prefetching [SDS00; KS02; LBM13], this chapter (to the best of our knowledge) is the first to use compiler-generated software threads for TLB prefetching. For the first time, this allows to accurately prefetch TLB entries for LDSes. Compared to the related compiler-generated software prefetchers, our solution is novel in how it issues fine-grained, timely prefetches into a shared resource (the TLB) in a scalable way without causing negative interference.

5.2 Target Architecture Template

The heterogeneous system targeted in this chapter combines two architecturally different processors in a single chip. As shown in Fig. 5.1, the HeSoC is composed of a general-purpose multi-core CPU (the host)

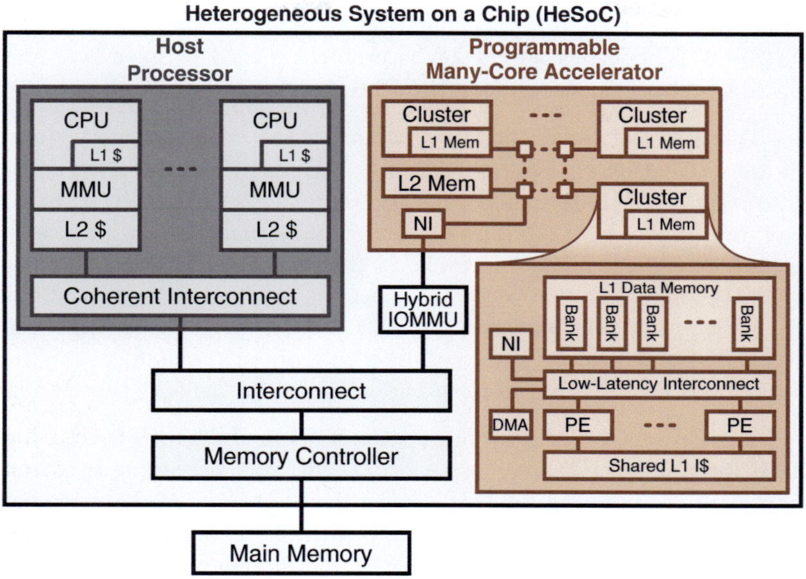

Figure 5.1: Template of the target architecture.

and a domain-specific PMCA. The host CPU features a cache-coherent memory hierarchy, runs a full-fledged operating system, and manages inputs and outputs of the HeSoC. The PMCA complements the host by offering high computational performance and efficiency for specific application domains.

The PMCA we consider uses a multi-cluster design [Mel+12; Kal14] to enable architectural scaling. In each cluster [Ros+17], multiple PEs share an L1 data SPM and an L1 instruction cache [Loi+18], both multi-banked (twice as many banks as PEs), through a low-latency logarithmic interconnect [Loi+15]. Multiple clusters are attached to the main network of the PMCA, through which they share an L2 SPM.

The off-chip main DRAM is physically shared by the host and the PMCA. To exploit data locality, both host and PMCA keep the most frequently accessed data in fast, local storage of their internal memory hierarchy. While the host relies on hardware-managed caches, the PMCA uses multi-channel, high-bandwidth DMA engines and double-buffering schemes to overlap data movement with computation on data in the L1 SPMs. The widely-adopted AXI protocol is used between host and PMCA and on the network in the PMCA.

The IOMMU allows the PMCA to share the virtual memory space of an application running on the host. It is a hybrid design [VMB17], consisting of a TLB of configurable size completely managed by software running on the PMCA. The TLB is fully associative and processes look-ups within a single clock cycle. In case of a TLB miss, the IOMMU stores the metadata in a hardware queue, responds with an error, drops the transaction, and processes the next one. In case of a TLB hit, the IOMMU translates the virtual address to a physical one, forwards the transaction through the master port, and processes the next transaction.

One PE of the PMCA manages the TLB in the IOMMU. Upon a TLB miss, it reads the metadata of the missing transaction from the hardware queue in the IOMMU, walks the page table of the offloaded process, replaces an older TLB entry with the new translation, and notifies the PE that encountered the miss, which then retries the memory access. As the memory access latency in page table walks dominates the miss handling latency, this software PTW has about the same latency as a dedicated hardware PTW [Vog+17].

PEs within a cluster execute in a SPMD fashion and share a multi-ported, multi-banked instruction cache [Loi+15]. They can exchange data with low latency and low congestion through the L1 data memory, which also offers an atomic *test-and-set* read-modify-write operation. A dedicated event unit within the cluster supports interrupts, barriers, and software-triggered events for low-overhead synchronization.

Each cluster includes a DMA engine optimized both in throughput and area for transfers from or to the cluster's tightly-coupled SPMs [Ros+14a]. It supports up to 16 outstanding AXI bursts with only minimal internal buffers thanks to the low-latency connection to the SPMs. Each PE has a private command interface on the DMA engine, which allows multiple PEs to simultaneously enqueue DMA transfers without the need for synchronization. The control unit of the DMA engine internally arbitrates between the per-PE command interfaces. PEs can enqueue coarse-grained transfer commands (up to 64 KiB), which are split up by the control unit into fine-grained bursts (up to 2 KiB) to meet alignment requirements and to facilitate time-multiplexing of downstream AXI resources. As soon as a coarse-grained transfer is complete, the DMA engine notifies the PE via the event unit.

5.3 Implementation

In this section, we detail our compiler-generated TLB prefetchers, which significantly reduce the rate of TLB misses (§ 5.3.1), our scalable multi-threaded TLB miss handlers, which keep the TLB miss handling latency constant for scalable parallel workloads (§ 5.3.2), and our hybrid-IOMMU-capable DMA engine, which can handle TLB misses without additional data buffers (§ 5.3.3).

5.3.1 Helper Thread Prefetching

As TLB miss handlers are dominated by memory latency, frequent TLB misses inevitably entail a large run time overhead. As a consequence, managing the TLB solely by reacting on misses is not sufficient. Instead, TLB entries could be set up ahead of the instant they are used in a prefetching manner.

The following observations motivate our prefetcher design:

- It shall not rely solely on run-time information (e.g., memory content, memory access patterns). This is the black-box approach taken by hardware prefetchers, which is not accurate for LDSes.
- It shall not rely solely on compile-time information (e.g., algorithms, data structures) because this neglects all dynamic information (e.g., data-dependent memory accesses, delays due to interference) required for timely prefetches.
- It shall be portable across applications. While software prefetches can be inserted manually, doing so effectively requires in-depth knowledge of the target platform and laborious analysis of the application. Prefetch insertion shall be fully automatic, not burdening developers.
- It shall exploit the cluster architecture of the PMCA, where tightly-coupled L1 SPM allows to share the state of PEs with low latency and little interference.

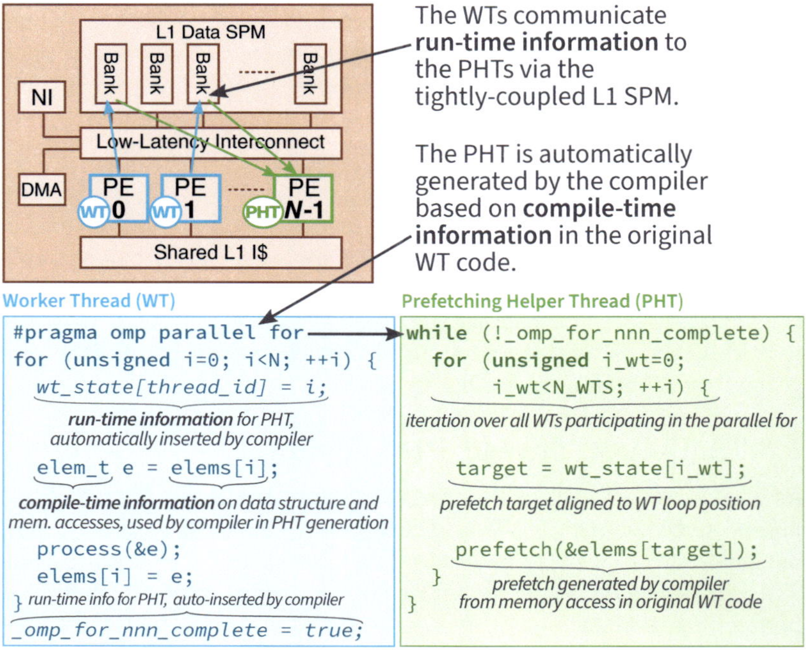

Figure 5.2: The concept of prefetching with closely-coupled helper threads using the example of a very simple parallel `for` loop.

- Its prefetching throughput shall be scalable at compile-time (because different programs have different memory access patterns and intensities) and at run-time (due to phase-based program behavior [She+03]).

Combined, these considerations led us to the concept of prefetching with closely-coupled helper threads shown in Fig. 5.2. Our execution model assumes that part of the PEs in a PMCA cluster are statically allocated to executing the original application workload. The workload is distributed among as many *Worker Threads (WTs)* according to the semantics of parallel programming models such as OpenMP [Mar+15]. The remaining PEs in the cluster are statically allocated to execute *Prefetching Helper Threads (PHTs)*, which our compiler automatically generates by stripping down the code of the WTs: The idea is to remove all statements that do not access SVM and do not determine the address or the occurrence of an SVM access, and to replace SVM accesses in the remaining code with a call to a prefetch method. Additionally, the compiler inserts store instructions to the L1 SPM into the WT to share its state of execution and load instructions into the PHT to read the execution state. The prefetch method does not modify the TLB itself; rather, it checks if a page is currently in the TLB and, if it is not, informs the standard TLB miss handlers (through the queue of TLB misses) that a TLB entry must be set up ahead of the moment when a worker thread actually requires the data on a page. Prefetching TLB accesses are described in more detail in § 5.3.1.

Prefetches are issued conditionally on the current position of the WT relative to the current position of the PHT. For example, a fixed window wherein prefetches are issued can be defined: Assume w_k is the position of WT k in a parallel loop and d and D are the minimum and maximum prefetching distances, respectively. Then the PHT has to make sure that the position of its next prefetch for WT k, p_k, fulfills $w_k + d \leq p_k \leq w_k + D$. If $p_k > w_k + D$, then the PHT is ahead of WT k by more than the maximum prefetching distance and the PHT will not issue a prefetch. If $p_k < w_k + d$, then the PHT is behind the minimum prefetching distance and the PHT will set p_k to a position inside the window. When p_k is inside the window, the PHT will prefetch at p_k and then increment p_k.

Compiler Algorithm to Generate PHTs

The compiler algorithm to generate a PHT from the code for a WT comprises two stages: The first stage recursively traverses the abstract syntax tree (AST) of the body compound of the WT. In a forward pass, a data dependency graph (DDG) for each SVM variable (i.e., a variable that dereferences a pointer to SVM, possibly through other variables and pointers) is constructed. In a backward pass, memory accesses to DDG leaf nodes are rewritten as prefetches and all statements that are not in the DDG of an SVM variable are removed. The second stage recursively traverses the modified AST and prunes redundant prefetches.

The forward pass is recursively invoked on an AST node and scope tuple, where the scope is a list of variables together with their DDG. It creates the PHT for the given AST node by constructing a DDG for each SVM variable and invoking the backward pass afterwards. The forward pass essentially distinguishes two classes of AST nodes: Declarations extend the scope of the subsequent nodes, and assignments modify the DDG of their left-hand side variable. Compounds establish a local scope, within which the children of the compound are first modified in forward order by the forward pass, then in backward order by the backward pass.

The backward pass is also invoked on an AST node and scope tuple. Based on the DDG of each variable, it rewrites dereferences of SVM pointers that are leaf nodes into prefetches and removes all statements that are not in the DDG of an SVM variable. It distinguishes three classes of AST nodes: Conditionals and loops add a control flow dependency to variables they reference. Declarations remove the declared DDG node from the scope (since the variable is undeclared before the declaration). Finally, assignments are either replaced by prefetches, left intact, or dropped completely, based on whether their left- and right-hand sides contain SVM variables.

Hardware Requirements

A prefetch load or store is slightly different from a regular memory access: upon a hit in the TLB, a prefetch transaction must not be forwarded downstream the memory hierarchy but instead be directly replied by the IOMMU as hit (with don't-care data in case of a load, since data returned by prefetch loads is ignored). A prefetch that misses in the TLB is not different from a regular miss to the hybrid IOMMU: it responds with a miss and drops the transaction. Conventional IOMMUs cannot support prefetches, since they lack the possibility to drop transactions and the masters using them lack the support for reacting to miss responses. Whether a load or store is a prefetch can be determined by a single bit sent with the request. Our implementation uses one bit in the AXI *user* field.

5.3.2 Multi-Threaded TLB Miss Handling

In the original implementation of the hybrid IOMMU we are using, TLB misses (only metadata) were enqueued by the IOMMU in a hardware queue [Vog+17]. This leftover from conventional IOMMUs presented a centralized bottleneck not required by the design, so we removed it and instead let PEs add an entry to a software queue located in the L1 data memory of their cluster upon a TLB miss. This atomic queue supports multiple parallel consumers and producers, and we implemented the atomicity with one enqueue mutex and one dequeue mutex based on the test-and-set functionality of the L1 memory.

For algorithms that make heavy use of SVM (especially those processing LDSes), a single miss handling thread (MHT) cannot cope with the rate at which WTs enqueue misses. In this case, the rate at which TLB misses are handled becomes the bottleneck. As an MHT is dominated by memory latency of the page table walking steps, the way to increase the miss handling rate is to let multiple MHTs work on different pages in parallel.

Two aspects are central for the design of the parallel MHTs: (1) given the sequence of misses in the queue, which MHT handles which miss, and (2) which MHT modifies which TLB entry.

For distributing misses among the MHTs, the simplest approach would be to let each MHT dequeue a miss, walk the page table, reconfigure a TLB entry, and wake up the PE that enqueued the miss. However, as two subsequent misses frequently go to the same page due to data locality (for an individual PE, e.g., with DMA bursts, but also for multiple PEs with shared data), this approach is not effective: Whenever a MHT dequeues a miss to a page that another MHT is already working on, it wastes run time and memory bandwidth on a redundant page table walk, and it wastes TLB capacity by setting up a redundant entry. Ideally, each MHT would dequeue all misses on the same page, walk the page table, and then wake up all PEs waiting for that page. However, this would require each MHT to traverse the entire miss queue (which can contain dozens of entries), locking both mutexes while it rearranges the queue without the misses to that page. This can take hundreds of clock cycles, during which no other PE can enqueue or dequeue misses.

Neither wasteful redundant miss handling nor making the miss queue a sequential bottleneck are acceptable, and our design avoids both: The MHTs share their state, i.e., which page each MHT is currently working on and which PEs it is going to wake up, through one word per MHT in the L1 data memory. When MHT A dequeues a miss, it first checks if another MHT is already working on the same page. If so, A tells the other MHT to also wake up the PE that caused the miss A just dequeued and dequeues another miss. If no other MHT works on the same page, A performs a prefetch memory access to the page to check whether the page has not been mapped since the miss. If the prefetch misses, A sets its state to that page and walks the page table. When A is done, it reads its state (which may have been updated in the meantime by other MHTs) and wakes up all assigned PEs.

Modifying a TLB entry takes two writes because virtual and physical page frame number together are longer than one data word. To avoid inconsistencies, the MHTs must thus ensure mutual exclusion when modifying a TLB entry. Any two different TLB slots are indepedent, though, so an MHT should not preclude another from simultaneously modifying a different slot. As both TLB levels are highly associative, MHTs have multiple options for the placement of each TLB entry. To make effective use of associativity, the MHTs should agree upon one replacement order per set. These three requirements can be met by using one atomic counter per TLB set, located in memory shared by all MHTs, which determines the index of the entry to be replaced next in a set. An MHT determines the set number from the virtual page frame number, increments the atomic counter of that set, and modifies the entry at the index returned by the counter. If the number of MHTs is comparable to the number of entries per set, the MHT must additionally lock the entry it modifies.

5.3.3 Hybrid-IOMMU-Capable DMA Engine

A hybrid IOMMU requires all masters that use it to be capable of tolerating TLB misses and keep track of which transactions missed. The DMA engine, however, was originally not designed to deal with error responses in a recoverable way and reported a transfer as complete as soon as it had received the final read or write response of the last burst, regardless of whether all bursts were successful or not. Thus, when a PE saw the completion of a transfer it started, it had no way to tell whether all data read or written were valid at the destination. To guarantee data integrity, all TLB entries required for the completion of a transfer (which can touch up to 17 4 KiB pages) had to be locked before the transfer could be programmed to the DMA engine and unlocked after it had completed. As the TLB is shared by multiple clusters, this limited the number of DMA transfers that could be enqueued at a given time and substantially reduced the effective data transfer bandwidth.

If the DMA engine can keep track of bursts that missed in the TLB and restart them after the miss has been handled, DMA transfers through the hybrid IOMMU can be much more efficient and scalable: An AXI burst may not cross a page, so each burst requires exactly one TLB entry at the instant its request arrives at the IOMMU. Requests of consecutive bursts can arrive back-to-back at the IOMMU, so multiple TLB entries need to be present only for a short time interval for an entire transfer spanning multiple pages to succeed.

To make the DMA engine compatible with the hybrid IOMMU, we designed and implemented a *retirement buffer* that keeps track of in-flight bursts. An entry in the buffer contains all metadata required to uniquely identify and reissue a burst: cluster-external and -internal address, length, AXI ID, DMA transfer ID, and whether it was a read or a write. When the AXI interface of the DMA engine sends a request, it adds a new entry to the retirement buffer, and when it receives the final response of a burst, it reports the success or failure of the burst with the responded AXI ID to the retirement buffer.

The retirement buffer must keep the order in which bursts were issued (because AXI bursts with the same ID are ordered) and must be able to complete bursts with different AXI IDs in any order. For these reasons, the retirement buffer can not be a simple FIFO queue. An alternative would be to have one FIFO queue per AXI ID, but this would waste hardware since every queue would need to have the capacity to store the maximum number of in-flight transfers.

Instead, our retirement buffer is a linked list implemented in hardware as shown in Fig. 5.3: The list entries are stored in a small register file that has as many words as the maximum number of in-flight transfers. Every word is wide enough to store the burst metadata mentioned above, the state of the entry (free, in-flight, failed, peeked, reissuable), and the index of the next burst entry. Additionally, the retirement

addresses		length	IDs		$\overline{R}$/W	state	next	
external	internal	[Byte]	AXI	DMA				
						Free		
0x2462 3...	(...)	(...)	5	2	0	Reissuable		
0x2468 0ff0	0x3 55c0	16	7	1	0	Failed		
0x2462 3...	(...)	(...)	6	4	1	Reissuable		
0x2497 1...	(...)	(...)	1	3	1	In Flight		←tail
0x2468 1000	0x3 55d0	240	0	1	0	Peeked		
0x2468 0ef0	0x3 54c0	256	4	1	0	Peeked		←head
						Free		

Figure 5.3: Organization of the burst retirement buffer.

buffer stores the index of the head, where order-preserving peek and pop operations start, and of the tail, where a new in-flight transfer gets enqueued.

The retirement buffer has three main interfaces: one to the AXI transfer unit of the DMA, one to the internal control unit of the DMA, and one to the PE control interface of the DMA. From the transfer unit, the retirement buffer receives commands to add a new in-flight transaction at the tail of the queue, to free a successful transaction, or to mark a transaction as failed. In the latter two cases, the buffer is traversed from the head, modifying the first non-free transaction that matches the given AXI ID.

To the DMA control unit, the retirement buffer reports the current number of in-flight and failed bursts and provides the metadata of the next reissuable burst. When at least one burst has failed, the control unit stops issuing new bursts from its queue and waits for all in-flight bursts to complete. Once there are no more in-flight bursts, the control unit reissues bursts as soon as they are reissuable until there are no more failed (and in-flight) bursts. As soon as this is the case, the control unit resumes regular operation by issuing bursts from its queue.

From the PE control interface, the cluster-external address of the first (ordered by request, not response) burst with state 'failed' can be read and bursts can be marked as reissuable. For this, a PE reads a DMA register to get the failing external address (or 0 if there is none).

Upon such a read, the retirement buffer marks all 'failed' bursts with the same page frame number as 'peeked' (so that the same page is not reported twice). Meanwhile, the PE determines the missing physical address and adds it to the TLB. When it is done, it writes the handled virtual address to the same DMA register, upon which the retirement buffer marks all 'failed' or 'peeked' bursts with the same page frame number as 'reissuable'. Bursts are then reissued by the control unit in the order of their original requests.

5.4 Results

In this section, we evaluate the performance of our SVM system implemented on an evaluation platform (§ 5.4.1) under various conditions (§ 5.4.2) to demonstrate its significant improvements over the state of the art and identify its limits (§ 5.4.3). Additionally, we discuss how our hybrid-IOMMU-capable DMA engine can save a vast amount of hardware buffers compared to conventional IOMMUs and standard DMA engines (§ 5.4.4).

5.4.1 Evaluation Platform

Our evaluation platform is based on the Xilinx Zynq-7045 SoC, which features a dual-core ARM Cortex-A9 CPU, which we use as host processor, and programmable logic, which we use to implement the cluster-based PMCA described in § 5.2. The PEs within a cluster share 8 KiB L1 instruction cache and 256 KiB tightly-coupled L1 data SPM, both split into 16 banks. Ideally, every PE can access one 32-bit word in the L1 SPM per cycle. Every cluster features a multi-channel DMA engine that is parametrized to have up to 8 AXI read or write bursts in flight at any time, enabling fast movement of data between L1 and L2 memory or shared DRAM. The PMCA is attached to the host as a memory-mapped device, interfaced through a kernel-level driver and a user-space runtime. The host and the PMCA share 1 GiB of DDR3 DRAM. The hybrid IOMMU features a two-level TLB: The L1 TLB features 32 entries, is fully associative, and translates addresses within a single cycle. The L2 TLB features 256 entries, is 8-way set associative, and translates addresses within up to 6 cycles. The IOMMU connects the PMCA to the ACP of the Zynq, allowing the PMCA to access the shared main memory coherent to the data caches of the host.

This platform enables us to study and evaluate the system-level integration of a PMCA in a HeSoC. Thus, we did not optimize the PMCA for implementation on the FPGA; the FPGA should be seen as an emulator instead of a fully-optimized accelerator. We adjusted the clock frequencies of the different components to obtain ratios similar to a real HeSoC with host and PMCA running at 2133 MHz and 500 MHz, respectively. The DDR3 DRAM is clocked at 533 MHz. Measuring an actual implementation rather than simulating models ensures all aspects and parameters of the evaluated system—including those we did not elaborate in detail in this chapter or might have overlooked—are correctly represented in the results.

5.4.2 Benchmark Description

To evaluate the performance of our SVM system under various conditions including identifying its limits, we have used two entirely different, configurable benchmark applications. They were obtained by extracting critical phases from real-world applications suitable for implementation on a HeSoC, and by parametrizing them over a large parameter space. They exhibit main-memory access patterns representative for various application domains.

Pointer Chasing (PC) This benchmark operates on graphs, stored as vertices linked by pointers. It is representative for wide variety of pointer-chasing applications from the graph processing domain [Guo+14]. Prominent examples include breadth-first or shortest path searching, clustering, and PageRank. Due to the irregular and data-dependent access pattern to shared memory and low locality between references, PC represents a worst-case scenario for a virtual memory subsystem. However, SVM is crucial to allow implementations of PC applications at reasonable effort and performance, because offloading a PC application to an accelerator without SVM requires modifying all pointers in a graph. In the benchmark, the host builds up a graph and stores its vertices in a single array in main memory. Every vertex holds the number of successors, a pointer to an array of successor vertex pointers, and a configurable amount of payload data. At the offload, the host passes a pointer to the vertex array and the number of vertices to the PMCA. On the PMCA, all WTs share the work of traversing the vertex array. For each vertex, a WT reads the number of successors and copies the payload data and successor pointers to a buffer in L1 SPM using DMA transfers. The WT then performs a configurable number of computation cycles on the payload and writes the payload to all successors in shared main memory again using DMA transfers.

Stream Processing (SP) This benchmark operates on a sequence of data, transferred from and to main memory in regularly strided blocks. It is representative for applications that work on streams of data, and examples range from simple one-dimensional filtering of audio data, over two- and three-dimensional image and video filters, to tensor operations in neural networks. In the benchmark, the host allocates one buffer of configurable size for both input and output (to maximize locality) and then passes the pointer to the buffer and the dimensions of the data blocks to the PMCA. On the PMCA, the WTs share the work of performing a configurable number of computation cycles on each block. Both input and output block are double-buffered in L1 SPM, so that compute and data transfer always overlap.

5.4.3 Benchmark Results

In the following plots, we compare the performance of our implementation and the prior state of the art (SoA) to an ideal IOMMU, which translates every address within a single cycle—an unbiased, although practically unreachable baseline. The SoA implementation is from [Vog+17], extended to multiple threads on the PMCA. For the relative performance on the y-axis, higher values are better. We evaluate different operational intensities on the x-axis by changing the number of computation cycles per data as described above. The operational intensity of an actual program depends both on the algorithm and the hardware executing it, and this sweep over a range of intensities characterizes our SVM implementation for a given memory access pattern but independent of a very specific program and processing architecture.

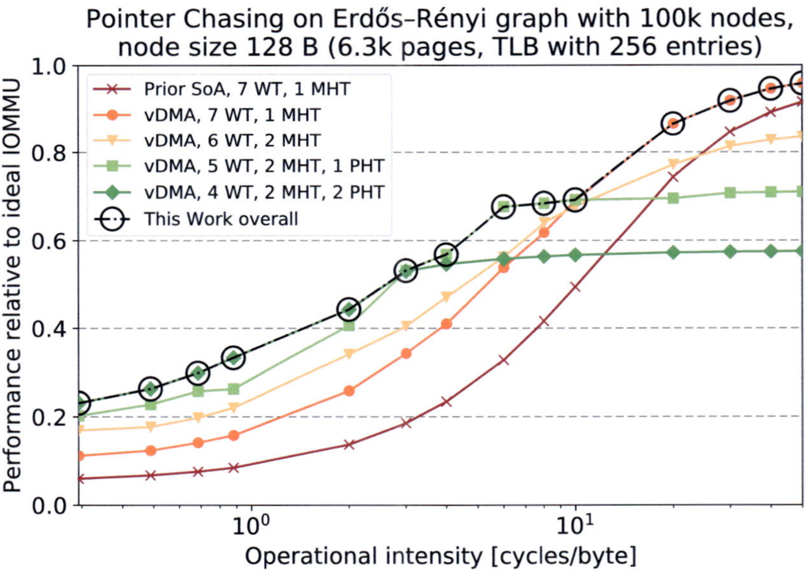

Figure 5.4: Pointer Chasing (PC) results for different operational intensities.

Pointer Chasing (PC) Fig. 5.4 shows the performance of PC normalized to an ideal IOMMU over different operational intensities in cycles per byte. For example, a single-precision floating-point implementation of the PageRank algorithm has an operational intensity of 1.2 cycle/B given a FPU with a divider and around 10 cycle/B for a reduced-precision fixed-point implementations if no FPU is available.

In the prior SoA, the DMA engine cannot handle TLB misses, so the software must ensure the TLB entries used by a DMA transfer are not replaced while that transfer is running. This locking is the bottleneck of the SoA implementation (first curve in legend order), and limits its performance to less than 50 % below 10 cycle/B. For very high operational intensities, the implementation becomes compute-bound and approaches ideal performance. Our hybrid-IOMMU-compatible DMA engine ('vDMA', all other curves) removes that bottleneck. The second curve is limited by the miss handling throughput of the single MHT for low operational intensities. Replacing one of the WT with another MHT (third curve) resolves this bottleneck. This is effective for low operational intensities, but the missing WT reduces performance in the compute-bound limit. Adding another MHT (not drawn) does not further improve performance because two MHTs are sufficient to handle the misses caused by six WTs. Instead, we replace one of the WTs by a PHT (fourth curve), which causes TLB entries to be set up ahead of the instant the WTs need them. This is very effective in the memory-bound case, increasing performance by another 20 to 30 %. The fourth curve is now prefetch-limited: the single PHT cannot always prefetch early enough, because the PHT itself needs to dereference pointers to determine prefetch targets. Any dereference that causes a TLB miss will block the PHT until the miss is resolved. Thus, replacing another WT with an MHT helps increasing performance in memory-bound cases by an additional 20 %.

Depending on the operational intensity, one of the configurations is optimal. However, as MHTs and PHTs can be inserted in software, e.g., at compile time based on profiling runs or even at run time for largely varying operational intensities, our work significantly improves performance for *all* operational intensities by making optimal use of PEs. The last curve shows the overall optimum configuration. For crucial operational intensities around 1 cycle/B, our work improves performance by 4 x compared to the SoA. For common intensities

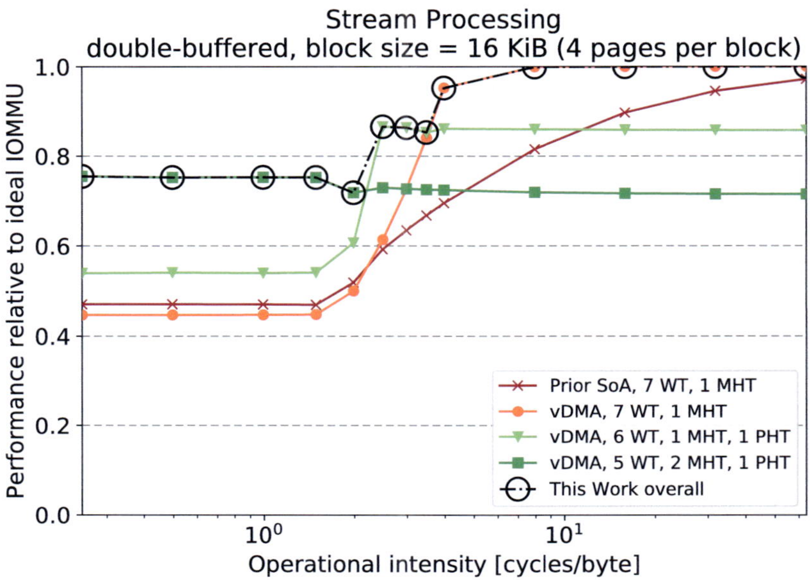

Figure 5.5: Stream Processing (SP) results for different operational intensities.

(arguably below 10 cycle/B), our work improves the SoA performance by at least 50 %.

Stream Processing (SP) Fig. 5.5 shows the performance of SP normalized to an ideal IOMMU over different operational intensities. For example, a one-dimensional FIR filter with N coefficients requires $N/2$ MACs per transferred data word (each word transferred once in, once out), and a matrix-matrix multiplication requires 1 MAC per transferred data word for large matrices. Assuming MACs on the data format are natively supported by the PMCA, the PMCA may compute tens, hundreds, or even thousands of MACs per cycle, depending on the number and architecture of its parallel PEs. Thus, stream processing kernel-architecture combinations may be found anywhere on the x-axis of Fig. 5.5.

In the prior SoA (first curve in legend order), a WT setting up a DMA transfer ensures that no TLB misses occur during the transfer by explicitly setting up TLB entries and locking them during the transfer. For memory-bound kernels, this is slightly more performant than handling misses by the hybrid-IOMMU-compatible DMA engine (second curve), because the latter stalls on every miss. (If that performance difference was larger, TLB entries could be set up in advance also for the vDMA. In contrast to the prior SoA, where locks on TLB entries had to be coded manually to avoid deadlocks, instructions for setting up TLB entries in advance for the vDMA can be inserted automatically at compile time.) For a range of more compute-intensive kernels, however, this locking is the bottleneck of the SoA implementation, and removing it improves performance by up to 35 %. When only few cycles are executed per transferred byte, performance is dominated by the memory latency in handling TLB misses. Adding another MHT (not drawn) does not change this, because only the input data stream requires one new page at a time. Instead, we replace one of the WTs with a PHT (third curve), which increases performance by 20 to 40 % in the memory-bound case. This configuration is limited by the *throughput* of the MHT, and because the PHT causes more than one page to be outstanding in the miss queue, there is now work for another MHT. Indeed, adding another MHT (fourth curve) increases performance by another 40 %, up to the point where it limited by the throughput of the PHT in the memory-bound case and by the five WTs in the compute-bound case. Adding another PHT might increase performance even further, but the current PHT generation algorithm does not support distributing the prefetches for a single memory access stream among two PHTs.

The last curve shows the overall optimum of all configurations of our work. Our work improves performance compared to the SoA by up to 60 % for memory-intense kernels, and reduces the overhead compared to an ideal IOMMU to below 25 % for *any* operational intensity. Our work also reduces the operational intensity at which that overhead is below 10 % to ca. 4 cycle/B. As the optimal configuration again can be selected at compile time or even changed at run time, our work significantly improves performance over the full spectrum of operational intensities also for very regular memory access patterns by replacing WTs with MHTs or PHTs when it improves overall performance.

5.4.4 Hardware Requirements of Hybrid-IOMMU-Capable DMA

Making the DMA engine compatible with the hybrid IOMMU not only improves performance compared to the SoA, it also dramatically reduces the amount of memory required to buffer DMA bursts that miss in the TLB. Our DMA engine is parametrized to have up to 8 AXI read or write bursts in flight at any time. Each burst can transfer up to 2 KiB. The total maximum amount of data in flight is 16 KiB, and a buffer of this size would be required to enable other masters to continue accessing SVM in the worst case scenario where 8 write bursts miss in the TLB. The retirement buffer in our DMA engine stores just the metadata of each burst: 32 bit for the virtual start address, 16 bit for the local start address, 3 bit for the ID, 8 bit for the length of the burst, and 3 bit for the status of the burst; less than 8 B in total. Thus, the retirement buffer requires just 64 B for the same TLB miss tolerance as the 16 KiB data buffer—a factor 256 less.

5.5 Summary

In this chapter, we presented and evaluated our scalable and efficient SVM solution for HeSoCs. It is based on a hybrid IOMMU and advances the state of the art in three important ways: First, compiler-generated PHTs proactively fill the TLB to minimize the rate of TLB misses. Second, a variable number of parallel PHTs handle TLB misses to scale the miss handling throughput with the demand. Third, a hybrid-IOMMU-capable DMA engine supports parallel burst DMA transfers to SVM without additional buffers. Compared to the state of the art, our work improves PMCA performance for memory-intensive kernels by up to 4× for irregular and by up to 60 % for regular memory access patterns. Compared to using data buffers to absorb bursts from DMA engines in a conventional IOMMU, our solution requires two orders of magnitude less memory and scales better, as it only stalls the missing DMA engine. In the future, we plan to explore compiler-generated PHTs for kernels that mandate speculative prefetching, improve per-thread miss handling throughput by supporting out-of-order page table walking, and avoid stalling the entire DMA on a TLB miss while maintaining memory order guarantees.

Chapter 6

Mixed-Data-Model Heterogeneous Compilation and OpenMP Offloading

Heterogeneous computers unite high versatility with high performance and energy efficiency by combining a general-purpose *host* processor with domain-specific PMCAs. The host manages input and output data as well as the application memory and *offloads* tasks that are highly parallel and/or domain-specific to one or multiple suitable accelerators [Nvi14; DKR18]. Due to the complexity of programming these systems, significant effort has been spent on developing programming models that retain high programmer productivity. A common way is to abstract the complexity through code annotations, indicating which code is to be offloaded, providing one unified code base. One de-facto standard programming model is OpenMP [MMG16].

OpenMP 4.0+ [Omp4.0] enables work to be offloaded from host to accelerators with the `target` directive and has been adopted for GPUs [Ant+16] and PMCAs in general [Mar+15]. Data is shared by copying from the *application memory*, which is managed by the host, to the *local memory* of the accelerator before the offload and back after the offload.

Application memory is growing rapidly: today 64-bit addresses are sufficient to handle data of hundreds of petabytes distributed over multiple nodes [Vaz+18], but when multiple exabytes of data need to be addressed, 128-bit host processors will be required to manage application memory. Accelerators, on the other hand, are designed to work mainly on data in their local memory, which inherently grows at a lower rate than total application memory. The same trend can be observed in heterogeneous SoCs, where 64-bit hosts are common today, although accelerator memory is within the 32-bit addressable range [Cot+15].

This growing disproportion raises the question whether there is a fundamental need for accelerators to increase their *data width* solely to share pointers with the host. For each accelerator core, doubling the data width at least doubles the size of most of its components – the frontend, the register file, the ALU (where the multiplier even grows quadratically), the load/store unit (LSU), and most internal buffers [ZB19]. Furthermore, it usually doubles the longest combinatorial path, requiring at least one additional pipeline stage to prevent a reduction of the maximum frequency. For every executed accelerator instruction that does not fully exploit the wider data path (doubling single instruction multiple data (SIMD) parallelism), performance per area and efficiency of the accelerator effectively decreases. As with most other properties of accelerators [HP17], it is thus desirable to design the data width to match the needs of the target domain. To achieve this in the long term, as the application memory continues to grow, mixed-data-width systems are required.

The challenge in mixed-data-width systems is to transform offloaded pointers and types that have a data-model-dependent size from wide host values to narrower accelerator values while preserving their semantics and incurring as little run-time overhead as possible. While this could be done manually, doing so is error-prone and requires to rewrite existing libraries and applications. Therefore, heterogeneous compilers need to support multiple *data models* to bridge the disproportionate data widths in heterogeneous systems. However, to date, no heterogeneous compiler practically supports accelerators with a data model that differs from that of the host. Additionally, minimal hardware support to let accelerators access addresses outside their native data width has not been explored.

Contributions. In this chapter, we address these challenges. To our knowledge, this chapter is the first to:

1. Design and implement a mixed-data-model (64+32-bit) heterogeneous compiler, including full support for OpenMP offloading.

2. Discuss the challenges and options for implementing mixed-data-model compilation in current versions of the two main compilers, GCC and LLVM.

3. Discuss novel hardware options for extended addressing and implement and evaluate a minimal, non-intrusive option that does not require modification of any core or ISA.

Outline. This chapter is structured as follows: After introducing the relevant background concepts in § 6.1, we explore the solution space to mixed-data-model OpenMP offloading in § 6.2, present our compiler solution in § 6.3, and describe the minimal accelerator hardware support for extended addresses in § 6.4. We show that our solution allows a 32-bit accelerator to transparently share memory with a 64-bit host at overheads below 0.7 % on PolyBench-ACC kernels in § 6.5. We compare to related work in § 6.6 and conclude in § 6.7.

6.1 Background

In this section, we introduce the heterogeneous compute and memory architecture targeted by this chapter (§ 6.1.1), OpenMP (§ 6.1.2), and data models (§ 6.1.3), and we discuss the state-of-the-art in offloading (§ 6.1.4) and heterogeneous compilation (§ 6.1.5).

6.1.1 Target Architecture

Fig. 6.1 shows the architectural template of heterogeneous computers we target in this chapter. The general-purpose host CPU is coupled to one or multiple PMCAs via an interconnect over which they share

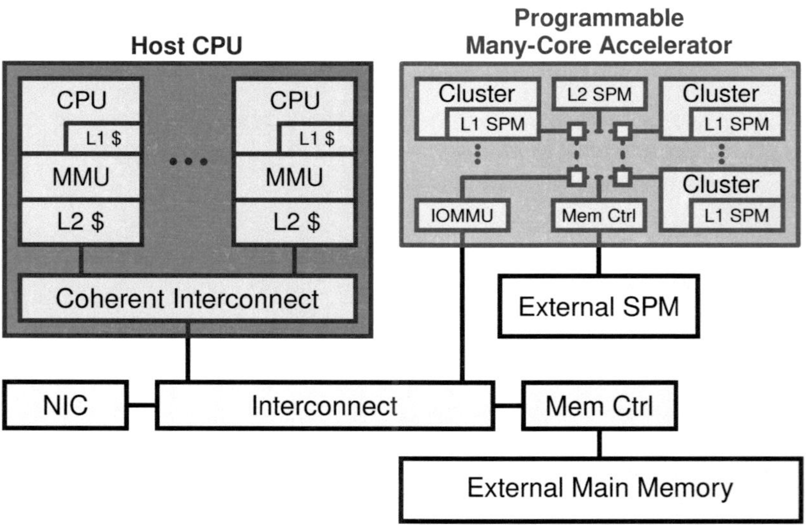

Figure 6.1: Architectural template of heterogeneous computers targeted by this chapter.

the external main memory and I/O peripherals, such as the network interface controller (NIC). The host CPU consists of one or more general-purpose application-class processing cores and has a memory hierarchy of virtually-addressed caches. The PMCAs consist of many minimal, domain-specific PEs, potentially grouped in clusters, have a memory hierarchy of physically-addressed, software-managed SPMs, and include an IOMMU to share the virtual memory space with the host. Host and PMCAs may implement different ISAs. There are many examples of such architectures in products ranging from HPC [Nvi14; Jou+17] over high-performance SoCs [DKR18] to low-power SoCs [Fru18; Tex19a] as well as in research [Cho+16; GH16; Kur+17; VMB18].

6.1.2 OpenMP and Offloading

OpenMP [Omp4.0] defines a target-agnostic API based on preprocessor directives that are translated by the compiler into calls to RTL functions. Since version 4.0, OpenMP supports *offloading* of computation to accelerators with the `target` directive and data sharing through the `map` directive. The `target` directive determines which code is compiled for the host, the accelerator, or both. GCC and LLVM implement this *heterogeneous compilation* in very different ways (§ 6.1.5), and we focus on how this impacts handling different data widths of host and accelerator.

The `map` clause of the `target` directive specifies data to be shared for each offloaded kernel. OpenMP's data sharing model is copy-based: The host copies data from its virtual memory space to a physically-contiguous memory section, which accelerators can access without participating in the virtual memory system of the host. This restricts `map` to data structures that do not contain pointers. However, extensions for SVM have been proposed and implemented [Mar+15; Vog+17; Kur+18a]. That generalized variant of `map` effectively reduces offloading to passing pointers to shared data to the accelerator.

Such true pointer sharing is essential for three aspects: First, it eliminates one level of copying (from the host to the device memory space (still DRAM) and back). Second, it allows the accelerator to transfer only the data it requires directly to its closest memory level. Third, it enables offloading of pointer-based data structures. OpenMP 5.0 introduced the `required` directive with the associated `unified_shared_memory` clause,

Data model	**Width (in bits) of**		
	`int`	`long`	**pointers**
ILP32	32	32	**32**
LLP64	32	32	**64**
LP64	32	64	**64**

Table 6.1: 32- and 64-bit data models common today.

which provides these pointer sharing semantics and makes `map` clauses on `target` constructs optional.

Our work supports both data-copy and pointer-passing offloading. When only copy-based offloading is required, simpler solutions could be found because the physically-contiguous memory section of the accelerator must inherently be addressable by the 32-bit accelerator.

6.1.3 Data Models

A *data model* defines the width of pointers and integer types that have a platform-dependent width. Table 6.1 lists 32- and 64-bit data models common today. In this chapter, we focus on pointers and discuss the challenges of offloading from a host with one data model to an accelerator with another in § 6.2.

6.1.4 Accelerator Address Space Restricted Offloading

There are already computers where accelerators have a narrower address width than the host [Red+18; Fru18; Jou+18; Cho+16]. To share addresses between host and accelerators on such computers without compiler support, different fallback options are being used. All these options restrict the address space of user-space applications on the host while keeping the OS in the native address space. First, host applications could be compiled for a different ISA that has a smaller address width but is compatible with the host ISA. For example, 64-bit ARMv8-A cores are user-space compatible with the 32-bit ARMv7-A ISA [Arm19a] and RV64 cores optionally implement an RV32 mode [Wat+19]. However, this is not possible for all ISAs. Second, host applications could be compiled for a different data model that has a smaller address width. For example, Intel introduced x32 for x86-64 [LAG11]. A major drawback of this option is that it requires changes to the compiler, the standard library, and the kernel, which are relatively complex to maintain for the limited benefits it offers [Lut18]. Third, some OSes, such as Linux, support restricting stack and heap addresses to a subset of the address space [Ker10]. However, none of these fallback options allow host applications to use the full 64-bit address space, so they do not solve the problem we address.

6.1.5 Heterogeneous Compilation: State of the Art

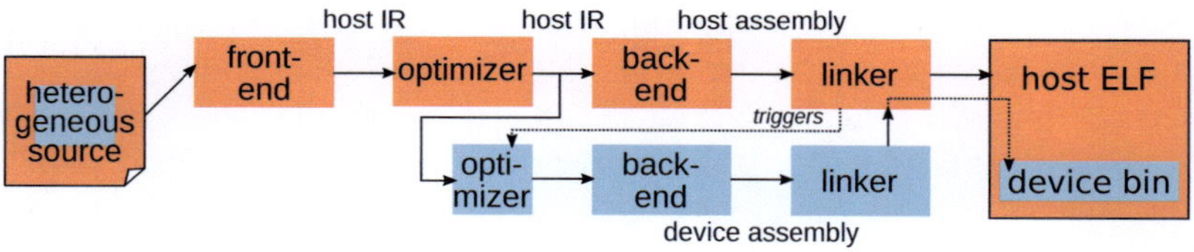

Figure 6.2: GCC implementation of OpenMP offloading. Red parts pertain to host compilation (top), blue parts pertain to accelerator/device compilaton (bottom). For device compilation, only a subset of optimization passes get executed.

GCC separates host and accelerator compiler to compile an application with OpenMP offloading, as shown in Fig. 6.2. The host compiler drives the compilation of a heterogeneous application and first lowers the source code to the GIMPLE IR. When the host compiler finds a `target` section, it creates a new outlined function. Next, in the expansion phase, the host compiler replaces OpenMP directives with calls to functions in the host `libgomp` RTL. Finally, after optimizing the GIMPLE IR and as part of LTO, the device compiler is invoked to transform GIMPLE IR to accelerator machine code [CM17].

In contrast to GCC, LLVM can natively compile for different targets and implements heterogeneous compilation as two separate compilations, as shown in Fig. 6.3. The necessary infrastructure was first presented in [Ant+16], the key feature being that the device compilation is largely independent from the host compilation, except for two points. First, the host compiler is responsible for annotating the parts of the source code

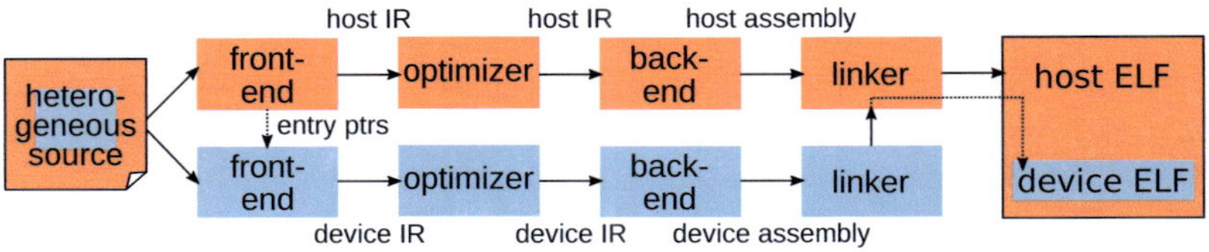

Figure 6.3: LLVM implementation of OpenMP offloading. Red parts pertain to host compilation (top), blue parts pertain to accelerator/device compilaton (bottom).

that are visible to the accelerator, and forwards this information to the device toolchain. Importantly, this is done before the host has assumed any data model upon the code. Second, the host link action depends on the completion of the accelerator compilation, such that the accelerator code can be linked into a fat host executable and linkable format (ELF) file. We will discuss the impact of the different approaches of GCC and LLVM in § 6.3.1.

6.2 Mixed-Data-Model Offloading

The central problem in mixed-data-model offloading is to overcome the difference in the width of pointers between host and accelerator. As a practical example, consider sharing 64-bit pointers from an LP64 host with an ILP32 device. The device's ILP32 data model defines pointers to be 32 bit wide, so host pointers cannot be used as function arguments on the device. The device ISA defines memory access instructions on 32-bit registers (and potentially immediates), but provides no way to access 64-bit addresses. Thus, even though data values wider than 32 bit can be shared with 32-bit devices, they cannot be used as pointers or, from a lower-level perspective, as memory addresses.

To access addresses wider than the native width, the device needs to provide minimal hardware support, which the compiler can use through builtin functions. We will describe the hardware implementation of these functions in § 6.4; but for now we assume there are two runtime functions the wide-address load `wide_load(uint64_t wideaddr)` and the wide-address store `wide_store(uint64_t wideaddr)` that take fixed-width integers (e.g., `uint64_t` in C), wide enough to represent the host pointers, as arguments and perform the load from or store to the given address.

6.2.1 Mixed-Data-Model *OpenMP* Offloading

To use host pointers as arguments to the extended load and store functions, they need to be converted to fixed-width integers in all OpenMP `target` code and the `map` clause. In C, this could be achieved by replacing all host pointer types with `uint`N`_t` and all reading or writing dereferences with calls to the load or store function, respectively. In C++, this could be achieved by defining a class that wraps a host pointer and overloading its dereference and assignment operator. When this transformation is left to the programmer, it is highly intrusive and requires changes to applications and libraries, which opposes the goal of transparent offloading. Moreover, this transformation is incompatible with copy-based OpenMP offloading for arrays because the array dimensions are stripped from the `map` argument.

The concept of passing host pointers as fixed-width integers and replacing their use in device code with calls to functions is nonetheless valid, but the transformation has to be performed by the compiler.

6.3 Mixed-Data-Model Compilation

In this section, we discuss options to implement mixed-data-model compilation based on the concept presented in § 6.2 in GCC and LLVM.

6.3.1 Feasibility in GCC and LLVM

GCC lowers the source code of host and accelerator to the same IR, which is determined by the host compiler (details in § 6.1.5). This implies that the data model of the host must be used also for the device, and since GCC treats all pointers uniform in this respect, that the data model of the host defines the width of all pointers. Unless GCC's approach to heterogeneous compilation is changed fundamentally and the GIMPLE IR can represent pointers of different width, mixed-data-model compilation is infeasible in GCC.

LLVM, on the other hand, separates compilation for host and accelerator as much as possible (details in § 6.1.5) including the use of multiple device-specific IR modules. Also, LLVM supports different *address spaces*, each of which can have its own width, and allows to assign pointers to address spaces. Address spaces are defined in the *data layout* string, and each heterogeneous target architecture can define its own data layout. This makes LLVM a natural choice for our mixed-data-model compiler.

6.3.2 Front-end or Optimizer?

The first question to address is where to implement the transformation of pointers. The choice is between front-end or optimizer, because by the time the code reaches the back-end, it has to be reduced to types and operations that the target supports natively. In the **front-end**, the transformation would traverse the AST of the application. It would identify each host pointer that is offloaded to the device, replace its type with a fixed-width one, and replace its use with a function call. The main drawbacks of this option are that matching all relevant patterns in the AST is difficult and that it has to be implemented specifically for each language that is to be supported. In the **optimizer**, the transformation would operate on the IR of the compiler. LLVM's IR is in SSA form, which allows for use-def chain traversals that are natural for replacing a pointer and all its uses. Also, the IR format is independent of the source language, so this option can be generalized to any language supported by the compiler. For these reasons, we implement our mixed-data-model compiler through optimizer passes.

6.3.3 Our Mixed-Data-Model Compiler in LLVM

To keep track of which pointers address values in the host and the accelerator memory, we assign them to separate host and accelerator *address spaces* (ASes). We refer to the accelerator AS as *device* AS to be aligned with common terminology. The *generic* AS defines the AS of pointers that are not explicitly assigned to an AS.

As shown in Fig. 6.4, we extend the compiler mid-end for the accelerator with two passes. The first pass assigns pointers from the generic AS to the host or the device AS. The second pass converts host pointers to fixed-width integers used to call the `wide_load()` and `wide_store()` builtin functions.

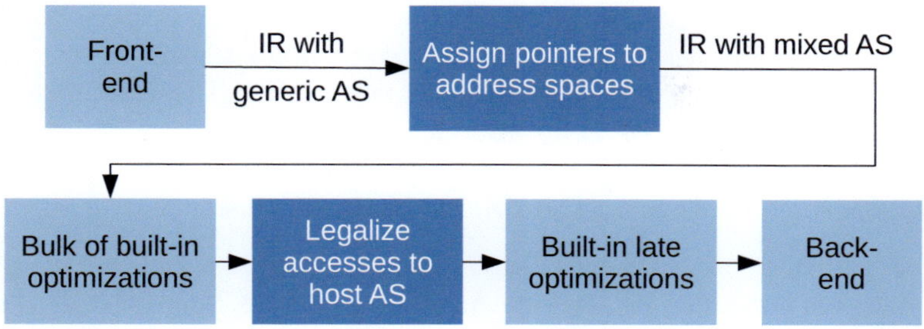

Figure 6.4: Flow chart of the extended (dark blocks) device compiler mid-end / optimizer passes.

Choosing the generic address space

As host and device compilation are separate in LLVM, the compiler assigns the native AS as the generic AS for compilation. However, using the native accelerator AS as generic AS leads to problems when the device also has to handle wider-than-native pointers: dereferencing a host pointer that is not assigned to the host AS on the device uses only the lower bytes of the address, which results in an illegal memory accesses. To ensure correctness, the compiler must guarantee that each pointer is assigned to an AS that is wide enough to cover the full addressable memory.

Compiler assignment of pointers to an AS is simple in many cases, e.g., pointers used in OpenMP offloading (`map` clause) are always in the host AS. In full generality, however, use-def chains can trace to a load from memory, where potential aliases make the identification of all possible values, and thus the decision between host and device AS, a difficult problem. To avoid this pitfall, we argue that the native device AS is no longer a suitable generic AS when considering mixed-data-width compilation.

We therefore switch to using the host AS as generic AS also on the device. As host pointers are assumed to be wider than accelerator pointers, the host AS is a superset of the device AS. This gives trivial guarantees for correctness as the generic AS is wide enough to represent any pointer. Having the fundamental correctness guarantees, the compiler can *optimize* for performance by assigning pointers to the device AS when it is guaranteed to preserve correctness.

For this reason, we use the host AS as generic AS and optimize as many pointers as possible into the device AS.

Assigning pointers to the device address space

Our solution for assigning pointers to address spaces has two stages. First, we introduce a `__device` qualifier and let Clang expand that to an `__attribute(address_space)` corresponding to the device AS, which propagates the AS assignment to the IR. The intention of this qualifier is that developers of accelerator libraries (e.g., device stdlib and OpenMP RTL) use it on pointer arguments and return values to give the compiler *anchor points* for device pointers. For instance, a `malloc` in accelerator SPM will return a device pointer, and data transfers to the local accelerator SPM will take a device pointer as `dst` argument. All stack allocations in the device code are automatically in the device AS. The second step is to propagate the AS from the *anchor points* during compilation, such that the burden is not put on the programmer.

```
EntryPtrs ← {}
for each module M do
    for each alloca A ∈ M do
        if A allocates a pointer of depth one then
            append A to EntryPtrs
        end if
    end for
end for
for each pointer P ∈ EntryPtrs do
    if HoldsOnlyDeviceASPointer(P) then
        NewP ← alloca of P in device AS
        ReplacePtr(P, NewP)
    end if
end for
```

Figure 6.5: Assign pointers to device AS if possible.

The AS assignments are propagated through pointers that only have dependencies to the known *anchor points*. For all pointers passed directly through use-def chains, LLVM does this implicitly. For pointers passed via memory (e.g., explicitly in C as pass-by-reference, when invoking OpenMP kernels, or due to a transformation), the `AddressSpaceAssigner` pass, shown in Fig. 6.5, ensures that the AS is properly propagated. We subsequently call the pointer stored in memory *inner pointer* and the pointer to it *outer pointer*. That is, *the outer pointer is used to pass the inner pointer via memory*. `AddressSpaceAssigner` first adds all stack-allocated outer pointers that store an inner pointer to the $EntryPtrs$ set (lines 1 to 8). On lines 9 to 14, the pass checks if each outer pointer $P \in EntryPtrs$ only ever holds device-AS inner pointers (Fig. 6.6). If it does, it replaces the generic-AS inner pointer in P with a device-AS inner pointer in a new outer pointer $NewP$, and recursively replaces all uses of P with $NewP$ (Fig. 6.7).

`HoldsOnlyDeviceASPointer` in Fig. 6.6 determines whether an outer pointer P always holds a device-AS inner pointer. For this, the pass matches all uses of P against conditions known to not change the AS of the inner pointer. The list of conditions constitutes the bulk of Fig. 6.6, but may not be complete. Since this is an optimization pass, the gist is to preserve correctness: If the conditions in the algorithm do not ensure that a pointer cannot be assigned a value outside the device AS, the AS migration is aborted on line 22. Additional conditions that preserve correctness could be added to further improve the optimization, but their omission does not compromise the correctness of Fig. 6.6.

`ReplacePtr` in Fig. 6.7 replaces each use of a pointer OP with a new pointer NP. We use it on line 12 of Fig. 6.5 to replace host-AS pointers with device-AS pointers. For each instruction U that depends on the old pointer OP, the algorithm first checks whether the use of OP in U can be replaced with the given new pointer NP (line 3). Such a replacement is not possible, e.g., if U is a call to an external function: Because LLVM IR is strongly typed – including ASes – the AS of every use of the argument within the external function would have to be changed, which is not possible if the function is not visible to the compiler. In this case, `ReplacePtr` casts the resulting pointer back to the original AS (lines 9 to 10), to remain compatible. If OP can be replaced (lines 4 to 7), U is cloned into NU, which uses the new pointer NP instead of OP. For dereference chains, `ReplacePtr` needs to replace the entire chain with the new AS. Due to strong typing, instructions that return pointers have their type changed when the AS is modified. This is detected on line 5 and triggers a recursive call on line 6. The base case

```
function HoldsOnlyDeviceASPointer(pointer P)
    ▷ Check if each use U does not set the inner pointer of P outside
      the device AS.
    for each use U of P do
        if U is ptrtoint or load from P then
            continue
        end if
        if U is cast or getelementptr then
            ▷ Casts and GEPs return a new pointer to recurse on.
            if HoldsOnlyDeviceASPointer(U) then
                continue
            end if
        end if
        ▷ Ensure that each inner pointer ever stored to P is in
          device AS.
        if U is store S to P and value of S is
              addrspacecast from device AS then
            continue
        end if
        ▷ Ensure that each function called with argument P does
          not modify the inner pointer of P.
        if U is direct call to function F and P is used
              read-only by F then
            continue
        end if
        ▷ If no match, inner pointer of P could be outside device
          AS.
        return false
    end for
    return true
end function
```

Figure 6.6: Determine whether an outer pointer P always holds an inner pointer in the device AS.

```
1: function REPLACEPTR(old pointer OP, new pointer NP)
2:     for each use U of OP do
3:         if can replace OP by NP in U then
4:             NU ← clone of U with OP replaced by NP
5:             if type of NU ≠ type of U then
6:                 REPLACEPTR(U, NU)
7:             end if
8:         else
9:             CI ← NP addrspacecast to type of OP
10:            replace use of OP in U with CI
11:        end if
12:    end for
13:    replace OP with NP
14: end function
```

Figure 6.7: Replace all uses of a pointer.

occurs when the type of NU is the same as the type for U, i.e., the instruction returns a non-pointer value. Once OP has been replaced in all uses, the original pointer OP is replaced by NP on line 13.

Following the AS assignment, the bulk of built-in optimization passes (e.g., canonicalization and loop optimizations) are executed on the resulting IR.

Legalizing accesses to host pointers

Before the late built-in optimizations, our second pass utilizes the AS assignments to legalize pointers before the IR is passed on to the back-end. *Legalization* is the process of converting generic IR types and operations to target-specific ones supported by the back-end. As all major ISAs today treat pointers like integers in terms of operations and register storage, pointers are just fixed-width integers in the back-end. The LLVM back-end can legalize operations on wider-than-native *integers* to multiple native instructions on multiple native registers. However, it can not generically legalize memory accesses to wider-than-native addresses.

This problem is solved by our `HostPointerLegalizer` pass, shown in Fig. 6.8, by utilizing the assigned address spaces. Wide host pointer

```
for each module M do
    for each load L ∈ M do
        if L loads from host AS then
            SZ ← size in bits of L
            IA ← integer address of L
            replace L with wide_load call of SZ to IA
        end if
    end for
    for each store S ∈ M do
        if S stores into host AS then
            SZ ← size in bits of S
            IA ← integer address of S
            replace S with wide_store call of SZ to IA
        end if
    end for
    for each addrspacecast A ∈ M do
        SA ← integer source address of A
        if SA in device AS then
            DA ← zero extension of SA
        else if SA in host AS then
            DA ← truncation of SA
        end if
        replace A with pointer from address DA
    end for
end for
```

Figure 6.8: Legalizing accesses to host pointers.

accesses are legalized by replacing them with calls to builtin functions `wide_load()` and `wide_store()` (as introduced in § 6.2, and to be defined in § 6.4), by replacing the wide pointers with fixed-width *integer* types that the back-end can already legalize. Specifically, the pass does the following for every module in device code. On lines 6 and 13, it replaces all loads from the host AS with calls to the `wide_load` function and all stores to the host AS with `wide_store` calls. Note that memory can additionally be accessed through intrinsic functions (e.g., *memcpy*), and the device RTL needs to provide implementations of these functions that can work with host-AS arguments. Finally, on line 23, the pass resolves AS casts from device to host AS by zero extension and from host to device AS by truncation of the source address. This truncation is lossless as Fig. 6.6 ensured that the pointer only ever holds inner pointers in the device AS.

Once pointers are legalized, late built-in optimizations (e.g., target specialization) can be applied before the IR is passed on to the target-specific back-end. Importantly, the *inline* pass is executed at this stage, to minimize the performance impact of the `wide_load()` and `wide_store()` functions.

6.4 HW Support for Extended Addressing

In the previous sections we left the `wide_load()` and `wide_store()` functions, which implement address-extended loads and stores, as black boxes. We will now define them. There are several options to implement this functionality in the underlying hardware. In this section, we present three options with decreasing degrees of intrusiveness, listed in Table 6.2, and implement the least intrusive option to show the generality of our solution and to upper-bound its overhead. For generality and because 64-bit addresses also induce a considerable amount of 64-bit *data* accesses, the examples discuss loading and storing 64-bit values. The reduction of the examples to 32-bit (and smaller) values with 64-bit addresses is trivial.

6.4.1 Additional, Wider Load & Store Instructions

The most intrusive option is to extend the ISA with custom load and store instructions that operate on paired registers. For example, a 32-bit

Option	Requires mod. ISA	Requires mod. Core	Instrs.	Cycles
Wider loads & stores	Y	Y	1	$L+1$
Adding CSRs	N	Y	4	$L+L'+2$
Mem.-mapped ext. reg.	N	N	6	$L+L'+4$

Table 6.2: Alternatives for accessing memory addresses wider than the data width of a core. The two right-most columns quantify the instructions and number of cycles of each alternative for loading (storing) a 64-bit value from (to) a 64-bit address with a 32-bit core. L is the latency of the first (or only) memory access, L' the latency of the subsequent access with an offset of 4 on the same base address. Modifications to the ISA also require modifications to the compiler backend.

ISA could be extended with instructions such as `ldd x0, 0(x2)` to load from `x3` (upper 32 address bits) and `x2` (lower 32 address bits) into the registers `x1` (upper 32 data bits) and `x0` (lower 32 address bits). Assuming a standard register file (RF) with two read and one write ports, each such load and store would take one extra cycle on top of the latency of the memory access, because the wider load needs to write the upper half of data to the RF and the wider store needs to read the upper half of address and data from the RF. Thus, this ISA extension allows a 32-bit core to access 64-bit addresses with one instruction and $L+1$ cycles, where L is the latency of the access. The compiler backend would need to be modified to know the double-register semantics of such instructions. This option requires logic to decode the additional instructions and a state register to control the address extension within the core but no additional register to hold the address extension.

6.4.2 Additional Control and Status Registers (CSRs)

As extending the ISA might not be possible, a less intrusive option is to add control and status registers (CSRs) to hold the part of an address that does not fit into registers. For example, one 32-bit CSR, which the LSU uses as upper 32 address bits, allows a 32-bit core to access 64-bit addresses. If the CSR is defined to clear on the next memory access and disable interrupts until the memory access (to prevent the address extension from corrupting memory accesses in interrupts), a 32-bit core can load a 64-bit value from a 64-bit address with the following four standard RISC-V[1] instructions:

```
csrrw  x0, csr_addr_ext, x3 // set upper half of address
                            // and disable interrupts
lw     x0, 0(x2)            // load lower half of data
                            // and reenable interrupts
csrrw  x0, csr_addr_ext, x3 // set upper half of address
                            // and disable interrupts
lw     x1, 4(x2)            // load upper half of data
                            // and reenable interrupts
```

where pre- and post-conditions on the registers are as in the last paragraph. The instructions setting the address extension CSR take one cycle each. The first `lw` might miss in the cache (latency L), while the second `lw` with an offset of four bytes to the same base address almost certainly hits (latency L'). Thus, this solution allows a 32-bit core with one additional 32-bit CSR to access a 64-bit address with 4 standard instructions and $L + L' + 2$ cycles.

6.4.3 Memory-Mapped External Register

The least intrusive option is to place an address extension register right outside the core and map it to the I/O address space of the core. For example, one 32-bit external register that extends the 32-bit address provided by the LSU of that core allows to access 64-bit addresses from an unmodified 32-bit core. Like the CSR, this register is defined to clear on the next memory access. A load with the same semantics as in the other examples can be performed with the following six standard RISC-V instructions:

```
csrrci x4, csr_status, 3    // disable interrupts
sw     x3, 0(mem_addr_ext)  // set upper (sic!) half of address
lw     x0, 0(x2)            // load lower half of data
sw     x3, 0(mem_addr_ext)  // set upper half of address
lw     x1, 4(x2)            // load upper half of data
csrrw  x0, csr_status, x4   // reenable interrupts
```

[1] Similar constructs are possible in other ISAs, we use RISC-V as a concrete example.

Like a CSR, a register directly after the core can generally be accessed in one cycle. Thus, this solution allows an entirely unmodified 32-bit core to access a 64-bit address with 6 standard instructions and $L + L' + 4$ cycles. The required extra hardware is one 32-bit register outside the core.

We implement this last option as it is the most generic and puts an upper bound on the overhead of our solution.

6.5 Evaluation

We show that our solution enables OpenMP offloading across data model boundaries with an average run-time overhead below 0.7 % compared to offloading restricted, native-accelerator-width addresses over a wide range of benchmarks.

6.5.1 Methodology

We implement our compiler in LLVM 8.0.0 [LLV19], and we use a custom version of the open-source HERO heterogeneous research platform [Kur+17; Kur+18b] to implement extended addressing as described in § 6.4.3. We use a 64-bit RISC-V Ariane core [ZB19] as host and a cluster from the PULP project [Ros+14b] with 8 32-bit RISC-V PEs [Gau+17], one DMA engine, and 256 KiB of L1 SPM in 16 banks that the PEs can access in a single cycle, as PMCA. Each PE has a memory-mapped external 32-bit register to extend addresses to 64 bits. All hardware is implemented in synthesizable hardware description language (HDL) and benchmarks are measured in cycle-accurate hardware simulation using Questa 10.7b [Men19].

We evaluate the seven kernels listed in Table 6.3. From the Polybench/ACC benchmark suite [Gra+12], `2mm`, `3mm`, `atax`, `bicg`, and `gemm` are linear algebra kernels, `conv2d` is part of the "stencil" domain, and `covar` is part of the "datamining" domain. Together, these commonly accelerated kernels span a wide range of memory access patterns and operational intensities. All matrices are stored in row-major arrays. Data is copied to and from accelerator L1 SPM with the DMA engine at the beginning and end of each offload phase, respectively. Computations of the accelerator thus exclusively use the L1 SPM. The accelerator PEs

Kernel	Parallelized computation	Complexity	
		space	comput.
`2mm`	$C_{i,j} = \sum_{k=1}^{N} \alpha A_{i,k} B_{k,j}$	$\mathcal{O}(N^2)$	$\mathcal{O}(N^3)$
`3mm`	$E = \text{2mm}(A, B) \rightarrow F = \text{2mm}(C, D) \rightarrow G = \text{2mm}(E, F)$	$\mathcal{O}(N^2)$	$\mathcal{O}(N^3)$
`atax`	$B_i = \sum_{j=1}^{N} A_{i,j} X_j \rightarrow Y_i = \sum_{j=1}^{N} A_{j,i} B_j$	$\mathcal{O}(N^2)$	$\mathcal{O}(N^2)$
`bicg`	$Q_i = \sum_{j=1}^{N} A_{i,j} P_j \rightarrow S_j = \sum_{i=1}^{N} R_i A_{i,j}$	$\mathcal{O}(N^2)$	$\mathcal{O}(N^2)$
`conv2d`	$B_{i,j} = \sum_{(k,l)=(-1,-1)}^{(1,1)} c_{k,l} A_{i+k,j+l}$	$\mathcal{O}(N^2)$	$\mathcal{O}(N^2)$
`covar`	$E_j = \alpha \sum_{i=1}^{M} D_{i,j};\ D_{i,j} -= E_j;\ S_{i,j} = S_{j,i} = \sum_{k=1}^{N} D_{k,i} D_{k,j}$	$\mathcal{O}(N^2)$	$\mathcal{O}(N^3)$
`gemm`	$C_{i,j} = \beta \left(\sum_{k=1}^{N} \alpha A_{i,k} B_{k,j} \right)$	$\mathcal{O}(N^2)$	$\mathcal{O}(N^3)$

Table 6.3: Evaluated kernels. Subscripts denote indices, uppercase letters are variables, and lowercase letters are constants. Arrows ($\rightarrow$) denote consecutive offloads. Semicolons (;) denote consecutive parallel phases within the same offload.

execute the computation in the second column of Table 6.3 in parallel. `3mm`, `atax`, and `bicg` are composed of consecutive offloads, denoted by arrows ($\rightarrow$) in the table, all other kernels consist of a single offload. All benchmarks are compiled with `-O3` but no specific optimization flags.

We measure the run time of each kernel in accelerator clock cycles, starting before the first DMA transfer of input data and stopping after the last DMA transfer of output data. In the *baseline*, the accelerator works exclusively with 32-bit addresses, i.e., the benchmarks are compiled for a 32-bit host and accelerator. We compare two implementations, where the benchmarks are compiled for a 64-bit host and a 32-bit accelerator with 64-bit generic AS, to the baseline: First, to analyze the performance impact of handling 64-bit pointers within the kernels on the 32-bit accelerator, we do not run the AS assignment pass. In other words, all accesses require 64-bit extension and take $L + L' + 4$ cycles, as described in § 6.4.3. Second, to analyze the efficiency of our solution, we run our full compiler including the AS assignment pass.

6.5.2 Benchmark Results

As the performance-critical part of each benchmark operates exclusively on device memory, which can be addressed with 32-bit pointers, our hypothesis is that the run-time overhead of our mixed-data-model compiler converges to zero with increasing data sizes. The evaluation is focused on small data sizes to analyze the effect on fine-grained offloading and the rate of convergence.

Fig. 6.9 shows the execution time of all benchmarks and both implementations relative to the baseline, where the accelerator works exclusively with native 32-bit addresses. For each benchmark, four bars represent different data sizes; for example, size 8 means that all matrices in a benchmark are 8×8 and vectors have length 8.

In the left part, where the accelerator has to work with 64-bit addresses also for local memory, the run time is multiplied by a factor of 1.4 to 5.8. For `3mm`, `2mm`, `gemm`, and `covar`, the relative run time converges to more than $4\times$. Those three kernels are dominated by computations and thus also by local memory accesses by the PEs using `wide_load`/`store`, and each access to a 32-bit word in L1 now takes 4 instead of 1 cycle. In addition, the `wide_*` memory accesses leave the compiler less freedom for scheduling memory accesses: it is currently not possible to define ordering constraints related to 64-bit addresses in a 32-bit compiler, so the order of every `wide_*` with respect to *any* other memory access needs to be preserved. The overhead is less pronounced for the other kernels, which are more balanced between data transfers through the DMA engine and memory accesses by the PEs. Nonetheless, the run-time overhead of using a 64-bit AS as generic AS for the device is clearly prohibitive, constituting the need for our AS assignment compiler pass.

In the right part of Fig. 6.9, where our compiler pass assigns as many device pointers to the 32-bit device AS as possible, the situation is completely different. Even for very small data sets (8×8 matrices and 8×1 vectors), the run-time overhead never exceeds 22 %. Even more importantly, the overhead rapidly converges to zero for all kernels, and already is below 0.7 % on average for still small data sets of size 64. This demonstrates the effectiveness of our AS assignment pass and proves that our compiler enables mixed-data-model offloading with negligible overheads in run time.

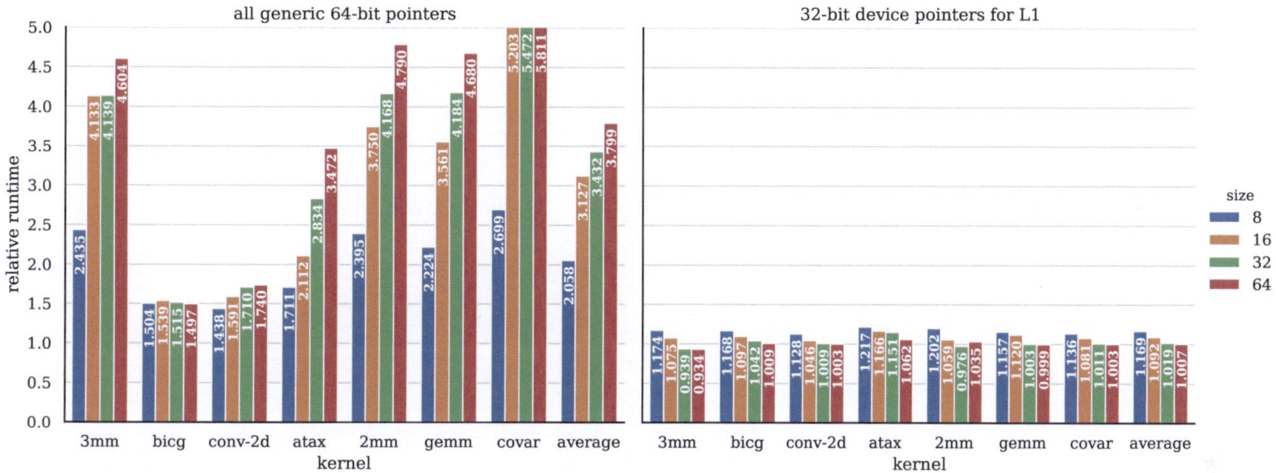

Figure 6.9: Execution time relative to using only 32-bit addresses without (left) and with (right) our compiler pass that assigns device pointers to the 32-bit device address space.

6.6 Related Work

To our knowledge, no computer today supports sharing data in the full host address space with accelerators that have a shorter data width. The concept of passing pointers as fixed-width integers (see § 6.2) could apply to such computers as well, given minimal hardware support for extended addressing (see § 6.4). However, we know no related work that includes these contributions (and consequently neither the compiler in § 6.3). In this section, we discuss how related works implement offloading by avoiding mixed data models.

Existing heterogeneous computers with mixed-width components (e.g., [Fru18; Red+18; Jou+18; Cho+16; Kur+17]) do not support OpenMP offloading or restrict the address space of offloaded applications to that of the accelerator (see § 6.1.4).

Most GPGPUs used today in heterogeneous computers can natively access 64-bit addresses [AMD18; Nvi19a]. GPGPUs, which implement double-precision floating-point arithmetics in hardware and are designed for SIMD (and SIMT) parallelism, naturally have a wide data path, so 64-bit addressing comes at very little additional cost. This also applies to CPU+GPU SoCs; for example, Nvidia's Tegra today fully supports offloading in 64-bit applications [Nvi16], although earlier versions restricted the address space of offloaded applications to 32 bit [Nvi15]. Many works address OpenMP offloading from 64-bit hosts to 64-bit GPUs with LLVM [Ant+16; Ber+17; Öze+18].

DSPs are an important class of accelerators that benefit much less from a 64-bit data path because they usually operate on sensor data, which does not exceed 32 bit in precision (although DSPs are also designed for SIMD parallelism). Today, even high-end DSPs are 32-bit machines [Tex19c] and the SoCs they are used in feature 32-bit host processors [Tex19a], even for driver assistance systems that include graphics accelerators [Tex19b]. Many of these fully-32-bit DSP SoCs support OpenMP offloading [Sto+13; Mit+14]. Other DSPs [Cod15; Cad18; Fru18] and DSP-like accelerators [Oh17; Cut16] that feature 32-bit very long instruction word (VLIW) ISAs are typically programmed through application-specific libraries such as OpenCV [Pul+12] and OpenVX [Khr19]. Provided a minimal OpenMP device RTL and support for the ISA in LLVM, our work could enable to use these accelerators as OpenMP offload targets in modern SoCs with 64-bit host processors.

Address spaces are also used in CUDA [Coo12] to define the memory location of functions and variables. Clang compiles CUDA code with address specifiers such as `__device__` to `addrspace` attributes in LLVM IR. In contrast to our work, no two CUDA address spaces overlap, so they are not used for accessing host memory from the device (or vice-versa) [Nvi19b]. Address spaces in LLVM can further define pointers to be non-integral, which has been used to implement "fat" pointers with hardware support for memory protection [Woo+14; Chi15].

OpenACC [CJ17] is a heterogeneous programming alternative to OpenMP. Its data model [Wol+17] is more generic than that of OpenMP but also supports accelerator-private memory. In Clang, OpenACC is being implemented by translating to OpenMP [DLV18], so our work naturally extends to OpenACC as far as implemented in Clang.

Different approaches enable automatic offloading to GPUs. Graphite-OpenCL [RAS10] first proposed static offloading of parallel loops to GPUs, relying on polyhedral analysis techniques to identify suitable subprograms. In the context of LLVM, several approaches use Polly [GGL12] as a foundation for automatic accelerator mapping. Examples are Kernel-Gen [Mik+14] which introduced a device focused approach only falling back to the host system if unavoidable, Damschen et al.'s approach [Dam+15] using a sophisticated client-server approach to orchestrate computations on Xeon Phi systems, and Polly-ACC [GH16] that introduces cross-kernel analysis to reduce overall data movement. To our understanding, all approaches target 64-bit devices and do not address offloading to devices with a data width that differs from that of the host. As our proposed concept for mixed-data-model compilation and offloading is not restricted to OpenMP but relies on generic IR analysis and transformations, it could apply to OpenCL and related frameworks as well.

Extended addressing has been implemented in processors for different purposes. In x86's Physical Address Extension (PAE) [Sha98], page table entries are 64-bit but the (virtual) addresses used by processors remain 32 bit. Using address translation to access a wider host address space from accelerators is theoretically possible, but maintaining a virtual address space that is different from that of the host is not trivial, so we prefer simple address extension. Similar to the first of our address extension options, [VT14] extended a 32-bit ISA with 64-bit load and store instructions and added a special-purpose register for address extension (whereas our first option does not require additional registers). They observe that SPEC CPU2006 uses less than 4 GiB memory and thus use 32-bit load/store instructions whenever possible. The authors have recently integrated that work into a proposed composite ISA [VBT19], which includes a 32-bit instruction subset. The focus of our work, in contrast, is to enable the first-class integration of 32-bit accelerators in a 64-bit addressed computer, and we design and implement compiler optimizations to do this in prevalent heterogeneous programming models and without restrictions or assumptions on the used memory space.

6.7 Summary

Our work extends prevalent programming models for heterogeneous computers (e.g., OpenMP, OpenACC) to computers with mixed data widths (e.g., 64-bit host and 32-bit accelerator) for the first time. We presented the general concept of mixed-data-model offloading in § 6.2 and designed and implemented an LLVM-based compiler to implement our solution fully transparently to the programmer in § 6.3. We discussed hardware support for extended addressing in § 6.4 and implemented the least intrusive variant to show the generality of our solution and upper-bound its overhead. Results on benchmarks from the PolyBench-ACC suite show that a 32-bit accelerator can transparently share memory with a 64-bit host at an average overhead below 0.7 % compared to 32-bit-only execution, enabling mixed-data-model systems to execute at near-native performance.

Chapter 7

A State-of-the-Art Open-Source Heterogeneous Research Platform

Heterogeneous integrated computing systems aim to combine general-purpose computing with domain-specific, efficient processing capabilities [Hor14; Zah17; DTH20]. Such computers integrate a general-purpose host processor with specialized programmable many-core accelerators (e.g., [DKR18; Ban+19; ABW20]). These systems are very complex and many challenges remain to be overcome to realize their full potential [HP19]. Central questions over the full stack of computing, from application programming over compilers and runtime libraries down to the accelerator microarchitecture, include: How to partition tasks between host and different accelerators? How to express that partitioning in programming languages, optimize it in the toolchain, or both? How to manage data sharing across host and accelerator, share address spaces and overcome the differences between cache-coherent memory subsystems, typically found on the host, and their non-coherent counterparts, which are typically found in accelerators? Which types and combinations of accelerators are optimal for a given domain?

Research on heterogeneous systems traditionally follows a double-track approach, where accelerators are developed in isolation [Reu+20; Gui+19], and their impact on system-level performance is estimated through analytical models and simulators [Uba+12; Pow+15]. Compared to using a prototype heterogeneous system, this approach has significant drawbacks: First, interactions between host, accelerators, the memory hierarchy, and peripherals are complex to model accurately, making accurate simulation orders of magnitude slower than running prototypes. Second, even full-system simulators model heterogeneous computers to a limited degree only [But+16]. For example, models of system-level interconnects or system memory management units (SMMUs), are missing or highly abstract and imprecise. Third, simulators that are not precisely calibrated and accuracy-validated against the simulated system are generally too inaccurate to provide reliable results, and full-system simulators are particularly unreliable [AS19]. A research platform that serves as a working prototype, on the other hand, enables collaborative and accurate architectural analysis and optimization [Lee+16]. To perform system-level research using standard benchmarks and real-world applications, the platform must additionally provide a software stack that includes an application programming interface and a complete compiler toolchain.

Existing research platforms do not meet these requirements in their entirety. Many provide a custom accelerator on programmable logic [Gra16; KHG20], and some even couple the accelerator to a host processor that runs an operating system [Man+20; Bal+20a]. HEROv1 [Kur+18c] provides software stack and compiler that enable the evaluation of real-world applications on a mixed-ISA computer, but it fundamentally restricts host and accelerator to use the same data model (e.g., 32-bit). Additionally, HEROv1's on-chip network and memory subsystem are restricted to simple architectures that cannot meet the demands of modern heterogeneous computers.

In this chapter, we make three main contributions:

1. We resolve the mentioned limitations and present HEROv2, an open-source research platform where an application-class 64-bit host can seamlessly share data with a 32-bit parallel programmable

accelerator. The latter is implemented on programmable logic and based on permissively licensed open-source RTL[1] components. The hardware components can be freely extended and modified and include a high-performance end-to-end on-chip network that can be fully customized to meet the memory demands of the accelerator and target application (§ 7.1.1). The platform also includes a complete heterogeneous compiler based on LLVM, which allows single-source single-binary development of heterogeneous applications with OpenMP 4.5 offloading (§ 7.1.2). The runtime libraries on the accelerator and driver on the host enable this offloading with little overhead (§ 7.1.3). A unified heterogeneous API enables productive programming while providing fine-grained control where necessary (§ 7.1.4). The complete software stack and tools are open source under a permissive license.

2. We demonstrate the capabilities of HEROv2 by using it to study four current research topics in heterogeneous computing and provide quantitative insights on the level of applications (§ 7.2.1), toolchains (§ 7.2.2), system architecture (§ 7.2.3), and accelerator architecture (§ 7.2.4).

3. We also leverage HEROv2 to design and evaluate a novel solution to one of the most pressing problems in heterogeneous computing: how to relieve the programmer of the burden of specializing an algorithm to the memory hierarchy of the accelerator (§ 7.2.2).

Section 2 describes HEROv2's hardware and software platform. Section 3 focuses on case studies. Furthermore, we compare with related work in § 7.3 and conclude in § 7.4.

7.1 Platform

HEROv2 consists of a complete heterogeneous hardware architecture (§ 7.1.1) as well as an end-to-end software stack including a toolchain and compilers (§ 7.1.2), operating system and runtime libraries (§ 7.1.3), and an application programming interface (§ 7.1.4).

[1]RTL = register-transfer level. We use the industry-standard SystemVerilog hardware description language.

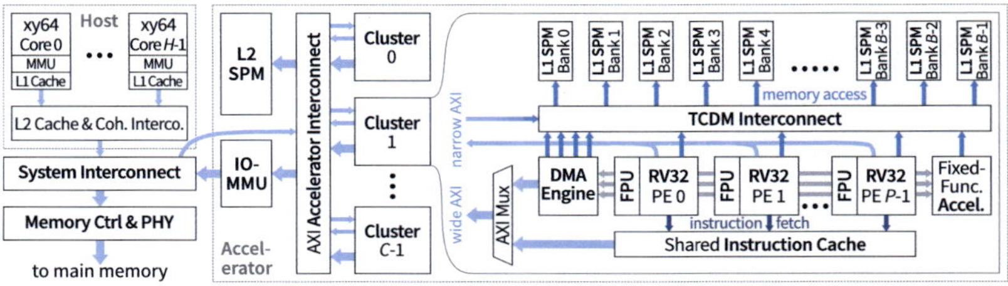

Figure 7.1: HEROv2's hardware architecture, which combines a general-purpose *host* processor (in the upper left corner) with a domain-specific programmable many-core *accelerator* (on the right side) so that data in main memory can be shared effectively (in the lower left corner).

7.1.1 Hardware Architecture

HEROv2's hardware architecture combines a general-purpose CPU running a full OS with a domain-specific programmable multi- or many-core accelerator. Fig. 7.1 gives an overview of the components and their connections. As many hardware components as possible are implemented on an FPGA (also called *programmable logic*) to make them configurable, modifiable, and extendable. All hardware modules mapped on the FPGA are available in synthesizable RTL logic under a permissive open-source license, which makes them fully analyzable and freely extensible and reusable. The vast majority of hardware components are silicon-proven, meaning they have been and will be used in ASIC tapeouts in many modern silicon technologies.

The host CPU is a hard-macro 64-bit ARMv8 Cortex-A multi-core on Xilinx' Zynq UltraScale+ family FPGAs or a soft-macro 64-bit RISC-V core (CVA6 architecture [ZB19]) on any UltraScale+ family FPGA. The general design principle of the host core is to maximize performance per area or power on mostly sequential workloads with a complex control flow. Each host core features a private L1 instruction cache, L1 data cache, and a MMU. All host cores are attached to a coherent interconnect and share an L2 data and instruction cache.

Host and accelerator share a main memory on a different die (such as HBM) or in a different package (such as DRAM) through the system interconnect, which can be coherent to the caches of the host. The memory hierarchy of the accelerator consists of software-managed SPMs.

To copy data into and out of the SPMs, the accelerator features DMA engines. To share the virtual address space of an application running on the host, each accelerator features a hybrid IOMMU (such as [Vog+17]). This IOMMU consists mainly of a TLB, which translates virtual user-space application addresses to physical memory addresses and supports a high degree of concurrency (e.g., tens of outstanding transactions from different masters). The TLB is managed by the accelerator itself, which handles TLB misses by walking the application page table managed by the host and filling the corresponding entries into the TLB. The IOMMU is called *hybrid* because it is managed in software, which allows the accelerator to efficiently share virtual address pointers with a minimum amount of hardware (e.g., for buffers).

The accelerator is composed of many minimal 32-bit RISC-V cores, which are organized into clusters of 4 to 16 cores for scalability. Different RISC-V core architectures are supported (see Table 7.1), and consequentially the specific ISA of the accelerator varies, but all accelerator cores support at least the `RV32IMA` ISA. The focus of the accelerator core architecture is to maximize the performance per area or power on computation-heavy workloads with a simple control flow. For this reason, the cores feature a single-issue in-order pipeline with 1 to 4 stages. To accelerate floating-point workloads, each core can be extended with a FPU, which is highly parametrizable: depending on the needs of the application, it can execute one double-precision (fp64) MAC, one or two single-precision (fp32) MACs, two to four half-precision (fp16) MACs, or four to eight quarter-precision (fp8) MACs in one clock cycle [Mac+21]. The optimal data width of an accelerator is one of multiple properties that depend on the target domain: for every executed accelerator instruction that does not exploit the full data width, its performance per area and per energy are suboptimal. As 32 bits suffice for many application domains, we expect 32-bit accelerator cores to remain useful.

To accelerate workloads that heavily rely on functions outside common integer or floating-point operations, the cores support custom bit-manipulation instructions, and the cluster can additionally be extended with fixed-function hardware processing engines [CSB18]. To maximize the utilization of the compute units, each accelerator core supports custom instructions to repeat a sequence of instructions multiple times without branches (so-called *hardware loops*) as well as custom instructions to implicitly increment the memory address on a load or store. Together, the custom instructions form the `Xpulpv2` RISC-V ISA extension [Gau+17].

Within each accelerator cluster, the cores have single-cycle access to a multi-banked, tightly-coupled L1 data SPM. A default banking factor of two allows any core to access any bank in any cycle with a low probability of contention for most applications. The cores can additionally access memory outside of the own cluster, including shared main memory, with a latency between a few (to other clusters) to hundreds of cycles (to main memory, depending on on-chip network and memory controller). A custom control and status register (CSR) allows each 32-bit core to load from and store to any 64-bit address [Kur+20b, § 5]. This CSR extends the native 32-bit address by 32 upper bit and is set automatically by the compiler (see § 7.1.2).

The cores fetch their instructions from an L1 instruction cache, which is shared by all cores in one cluster. To reduce the pressure on the shared instruction cache during loops, each core additionally contains an L0 instruction cache holding up to eight compressed instructions.

Finally, each accelerator cluster features a DMA engine, which can address the full 64-bit memory space, supports unified virtual memory through the hybrid IOMMU [Kur+18d], can transfer up to 1024 bit per clock cycle in and out of the cluster (full duplex), and can have tens of transactions, each consisting of tens of data beats, outstanding at any time. This DMA engine allows to transfer data in high-bandwidth bursts while the accelerator cores compute on data in local memory. If an application allows issuing sufficiently many or long bursts, the DMA engine allows tolerating a latency of hundreds of cycles between main memory and accelerator, which is crucial to support the ongoing trend of deeper and non-uniform memory hierarchies.

Multiple accelerator clusters are interconnected with two non-coherent networks implementing the industry-standard AXI protocol: a wide one for high-bandwidth DMA transfers and a narrow one for low-latency accesses by cores [Kur+22]. A high-bandwidth on-chip SRAM controller connects the L2 SPM, which is shared by all clusters, to the accelerator interconnect. The hybrid IOMMU connects the accelerator to the host. The IOMMU is either directly attached to a non-coherent system interconnect or via a bridge [Cav+20] as I/O-coherent request node to a coherent system interconnect. To program and control the accelerator from the host, it is additionally connected as I/O-coherent slave node to the system interconnect.

Such RISC-V-based MIMD/MPMD-capable accelerators can be seen as a functional superset of accelerators with a more specialized ISA and execution model (e.g., PTX or RDNA and SIMD/SIMT for GPGPUs). While this RISC-V accelerator might not reach the performance of a state-of-the-art commercial accelerator, it enables the community to develop solutions for many problems that are not exclusively caused by the performance of a specific accelerator. Internally, any company can replace the RTL code of the provided RISC-V accelerator with their proprietary code to obtain cycle-accurate results of their product within a heterogeneous system without having to disclose any trade secrets.

7.1.2 Toolchain and Compilers

HEROv2's heterogeneous hardware requires toolchain support to enable the development of applications for the platform in an efficient and productive way. HEROv2 provides such a heterogeneous toolchain, based on LLVM 12[2], which provides efficient support for heterogeneous compilation based on OpenMP. This enables the seamless co-integration of compute-focused accelerator kernels and control-focused host code into a unified application, including target-specific compilation and optimizations. Additionally, the different data widths of the system (64-bit host and 32-bit accelerator) are supported by LLVM's address space implementation, which provides the compiler with the means to express pointers of varying width. An overview of HEROv2's toolchain is shown in Fig. 7.2.

The toolchain flow starts by compiling OpenMP-annotated heterogeneous source code, as shown at the left of the figure. OpenMP describes heterogeneity and parallelism through `#pragmas` and leaves the transformation to parallel code to the compiler. Annotating a piece of code with `#pragma omp target` directs the toolchain to compile the code both for the host[3] and the accelerator. We refer to these regions

[2]We periodically update to the latest LLVM version. The effort for an update is usually on the order of a week for a full-time engineer.

[3]By OpenMP's specification, the runtime decides during execution time if a target region is executed on the host or the accelerator, but in the case of HEROv2, the latter is always the case.

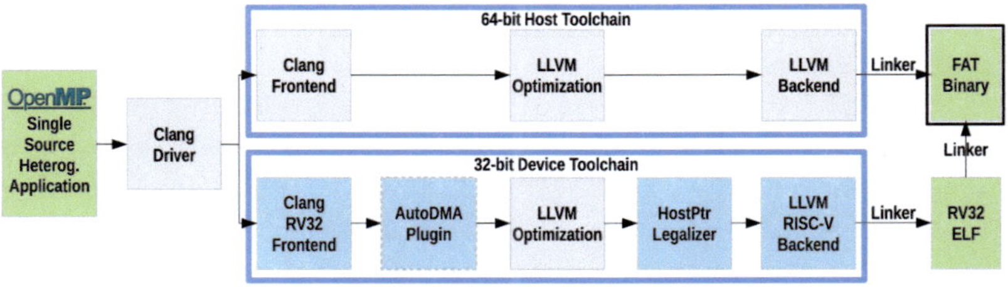

Figure 7.2: Overview of HEROv2's heterogeneous toolchain and compilers.

as *target regions*. The respective host and device toolchains are thereby invoked by the *Clang driver*, as shown in Fig. 7.2, which transform the source code into an object file for each architecture. The Clang driver triggers the device linker for the device object file, creating a RISC-V ELF file, whereafter the host linker is triggered. The host linker first links the host object files into an ELF file and then embeds the device ELF as an object inside the host ELF, creating a *FAT binary*. This allows the OpenMP runtime to load the device ELF into accelerator memory at runtime.

HEROv2 uses an off-the-shelf LLVM-based 64-bit toolchain for the host and a custom LLVM-based 32-bit RISC-V toolchain for the device. Both toolchains are marked in Fig. 7.2, where the components that include customizations are highlighted in blue tones, and aim to provide interoperability between host and accelerators, ease of programming, and support for ISA extensions.

Interoperability between Host and Accelerators

Pointers in C/C++, as well as in the LLVM IR, have a fixed width: the *data width* of the target processor. A 32-bit data width of an accelerator therefore implies that 64-bit pointers from the host will be truncated. To allow 64-bit host pointers to be correctly represented, an additional 64-bit *address space* is defined in HEROv2's accelerator compiler. We refer to the two address spaces as the 32-bit *native address space* and the 64-bit *host address space*. Address space support is a built-in LLVM feature, and has been previously used, e.g., to separate pointers to global and shared memory in CUDA. In such cases, however, pointers are annotated by the programmer, e.g., `__shared__` in CUDA, and all pointers typically have the same width.

To address mixed-data-width compilation in HEROv2, the *Clang frontend* has been extended to generate LLVM IR with automatically assigned address spaces. We adopt the techniques of [Kur+20b], where OpenMP offloading entry points are used to infer that pointers passed to a device kernel from the host are 64-bit wide. The uses of such pointers are then tracked throughout the application, such that any pointer that *cannot* be guaranteed to never hold a 64-bit host address is *promoted* to the host address space. Any pointer that is guaranteed to only hold 32-bit pointers is kept in the native address space. Additional control is handed to the programmer through `__device` pointer decorations, to enforce a pointer to belong to the native address space, if the compiler could not guarantee it to be correct. As part of machine code generation, any data types and operations that are not natively supported by the underlying hardware and/or application binary interface (ABI) must be *legalized*. As pointer semantics are dropped in LLVM backends (i.e., pointers are treated as integers), the backend is able to implicitly legalize arithmetic operations on 64-bit pointers. However, the backend does not support the legalization of wider-than-native load and store operations. HEROv2's RISC-V compiler has therefore been extended with a custom *host pointer legalizer* pass right before the optimized code is passed to the RISC-V backend for machine code generation. This pass identifies all load and store operations on addresses in the host address space and implements them using the address extension CSR.

Ease of Programming and Code Portability

An important aspect for code portability and ease of programming is the automatic optimization of code for the memory hierarchy of a computer. HEROv2's accelerators use software-managed SPMs, which are refilled using DMA engines. This means software must explicitly orchestrate any data movements between shared main memory and fast local memory. As OpenMP does not provide any mechanisms to tile data structures and move tiles with DMA transfers, programmers need to manually rewrite their code to perform well on SPM-based accelerators. HEROv2's DMA API is unified over all accelerators, but the initial tiling of an application is nonetheless a significant effort and reduces code portability outside HEROv2.

To reduce this effort and improve code portability, HEROv2's device compiler provides an optional *AutoDMA* plugin that automatically analyzes source code to identify memory regions that are suitable for staging through SPMs and transforms the code to automatically program the DMA engine without any programmer intervention. The AutoDMA plugin is also able to perform loop tiling to extract segments of code whose memory footprint is small enough to fit in the local memory. The AutoDMA plugin is an extension of *HePREM* [FBM20], originally envisioned for transforming real-time GPU code to be less sensitive to memory interference. This was achieved by transforming GPU kernels into a series of *load*, *execute*, and *store* phases, with explicit synchronization points between them. These three phases are well aligned with accelerators based on software-managed SPMs. In contrast to *HePREM*, which targets DMA-less GPU systems, *AutoDMA* generates DMA API calls instead of moving data using load and store instructions. Additionally, synchronization has been minimized to improve performance. The resulting *AutoDMA* plugin provides an optional way to achieve performance on HEROv2 without the need for manual tiling and DMA management code.

Support for ISA Extensions

The device compiler backend of HEROv2 has been extended to support the `Xpulpv2` ISA extension [Gau+17]. This includes the automatic detection and insertion of *hardware loop* instructions, automatic optimization to generate *post-increment* load and store instructions, as well as pattern matching to emit *multiply-accumulate* instructions, outlined in § 7.1.1. To the best of our knowledge, this is the first time custom instructions have been implemented for RISC-V in LLVM, and a full-system performance evaluation is shown in the case study in § 7.2.4.

In summary, HEROv2's heterogeneous toolchain provides a de-facto standard and heterogeneous-by-design programming model via OpenMP, which is fully supported by its LLVM-based compiler. This provides seamless, end-to-end single-source-to-heterogeneous-binary compilation. The device compiler has been significantly extended to support performance, ease of programming, and code portability: first, through the minimization of expensive wider-than-native load and store operations in a mixed data-model setting; second, through the support for automatic tiling and DMA management through *AutoDMA*; and third, through automatic code generation targeting the performance-oriented ISA extensions supported by the underlying hardware.

7.1.3 Runtime Libraries and Operating System Support

HEROv2's runtime software stack is designed to seamlessly integrate the accelerators into the OS running on the host and allow for transparent accelerator programming with OpenMP 4.5 offloading [Omp4.5] and unified virtual memory compliant with HSA specifications [Hwu16]. An overview of the runtime stack is shown in Fig. 7.3. This section discusses the layers below the API, which is discussed separately in § 7.1.4.

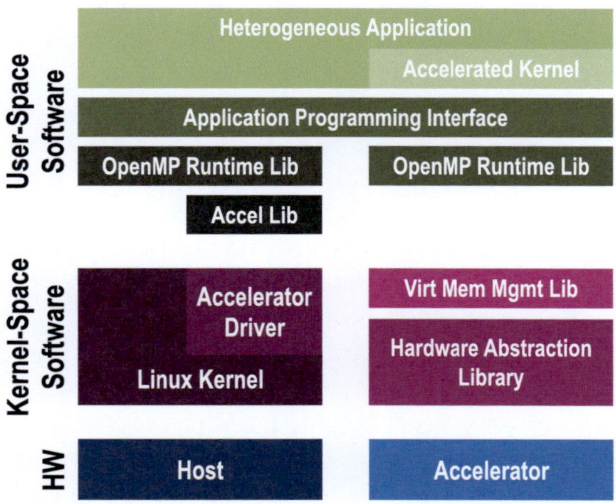

Figure 7.3: HEROv2's runtime stack, which seamlessly integrates *accelerators* (with their runtime stack on the right) into the OS running on the *host* (runtime stack on the left) to enable heterogeneous applications with transparent offloading of accelerated kernels (at the top).

A heterogeneous application starts executing on the host. When the host encounters a `#pragma omp target` directive, it offloads the code within the `target` region to the specified (or default) accelerator (called *device* in OpenMP terminology). To this end, the host OpenMP runtime library invokes the device-specific runtime plugin. The plugin passes a pointer to the offloaded code and data to a hardware mailbox in the device, thereby starting execution on the device. The first core of the first cluster of the device runs an *offload manager*. It is woken by an interrupt from the hardware mailbox and starts executing the offloaded function. All data items inside the `map` clause become available to the device: When unified virtual memory is enabled, pointers are passed unmodified, no data is copied, and the device is given read-only access to the user-space page table of the application on the host. Otherwise, the host copies data to a physically contiguous memory region in main memory and changes the pointers before passing them to the device. By design, offloading does *not* copy data to the SPMs of the device. There are two main reasons for this: First, HEROv2's accelerator model aims at accelerating kernels that take at least ten thousand cycles to execute. Thus, the offloading model is relatively coarse grained and the `map`ped data in its entirety in general does not fit into the local memory of the device. Second, OpenMP's `map` clauses cannot express tiling, yet flexible tiling is essential for efficient execution on device-local memory.

Inside the offloaded region, execution starts on the first core of the first cluster. When that core encounters a `#pragma omp teams` directive, it forks execution to multiple clusters, and the cluster master core (i.e., the first core of each cluster) starts executing the region. When a cluster master core encounters a `#pragma omp parallel` directive, it forks execution to multiple cores of its cluster. Inside parallel regions, all OpenMP worksharing, datasharing, and synchronization constructs are available, allowing for effective parallel programming following OpenMP's standard paradigm. The OpenMP device runtime library implements the functions emitted by the OpenMP code generation pass of the compiler (whose function names start with `__kmpc_`) by calling into the accelerator-specific hardware abstraction library (HAL).

The VMM library allows the accelerator to share the virtual address space of a user-space application running on the host (concept of [Vog+17]). After the host has set up entries in the IOMMU that allow the accelerator to access the page table, the VMM library provides functions to translate any valid virtual address to a physical address and set up a corresponding translation entry in the hybrid IOMMU. Application programmers usually do not notice this: The compiler generates the correct instructions for accessing pointers outside the native (32-bit physical) address space of the accelerator. In the common case, such accesses hit in the TLB of the IOMMU and incur an overhead of only three cycles per remote memory access [Kur+20b]. When an access misses in the TLB, the core either invokes the VMM library itself to add an entry to the IOMMU, or it lets a dedicated core handle the misses. The latter is preferable for pointer-based applications, and miss handling can be configured per offload through custom options to the `target` region. The implementation of the VMM library is specific to the virtual memory system of the host (e.g., ARM VMSAv8-64 or RISC-V Sv39 or Sv48).

The HAL on the accelerator provides functions for forking parallel execution, identifying and synchronizing cores, putting cores to sleep and waking them up, controlling the DMA engine, and communicating between clusters and with the host through the mailbox. The HAL is implemented using low-level hardware-specific primitives, such as writing memory-mapped registers and setting bits in CSRs.

The OS device driver and the accompanying user-space accelerator library on the host implement the accelerator-specific functionality for offloading to and communicating with the accelerator from the host. This includes identifying the accelerator in the device tree, resetting, initializing, and programming it, and making the page table of the user-space process readable for the accelerator.

In summary, HEROv2's modular runtime stack supports different hosts and accelerators while reusing large parts of the code base, combining flexibility with accelerator-specific specialization. On the accelerator, all runtime libraries are linked into the offloaded application, and LTO minimizes the overhead of the multiple layers. On the host, system calls are required to trigger and conclude an offload, but the overhead of that is negligible due to HEROv2's coarse-grained offloading model.

7.1.4 Application Programming Interface

The application level is the most important from the perspective of end users and application developers. HEROv2's toolchain and runtime software provide the means to make effective use of accelerators, but without a properly designed API applications on heterogeneous computers remain too complex to program in most cases. Porting an application to make efficient use of the software-managed memory of an accelerator involves tiling data and scheduling data transfers, which are difficult tasks in general. An API alone cannot solve this problem, but it can make the work of the application programmer portable over different accelerators and substantially easier by abstracting the intricacies of the hardware away. The design goal is to provide an interface that is unified over all supported accelerators together with an implementation that is optimized and verified for each accelerator individually. HEROv2's API complements the OpenMP API for offloading and parallel programming (§ 7.1.3) and the accelerator-specific compiler (§ 7.1.2), which optimizes the compute part of an application for the target accelerator.

HEROv2's API has three main categories of functionality: memory management for the different SPM levels, data transfers between SPMs and main memory, and performance measurements. All functions are thread-safe and can thus be used inside and outside parallel regions.

To manage the heap memory of the accelerator, there are three functions for each SPM level: `hero_lN_capacity` returns the amount of currently available heap memory at SPM level N. This function is often used at the beginning of a tiling region to calculate the tile sizes. `hero_lN_malloc` and `hero_lN_free` implement POSIX' memory allocation and freeing functions [POSIX.1-2017] for SPM level N. The implementation uses a deterministic constant-complexity memory allocator [Her14; Kir20], ensures mutual exclusion among all affected cores (e.g., within the same cluster for L1 SPM) through RISC-V atomic operations, and can detect heap overflows with a canary mechanism. The alignment and minimum allocation granule is 8 B.

To transfer data between SPMs and main memory, HEROv2 provides multiple functions with the semantics of POSIX' `memcpy` [POSIX.1-2017]. Those functions are organized in three dimensions: direction (device-to-host or host-to-device), synchronicity (blocking or asynchronous), and transfer dimensionality (1D, 2D, etc.). The direction has to be distinguished in the function signature because pointers and addresses in the host-managed main memory are of a different width and address space than device-internal pointers: in `hero_memcpy_host2dev_*` functions, the `src` pointer is in the host address space and the `dst` pointer in the device address space, and vice-versa for the `hero_memcpy_dev2host_*` functions.

The synchronicity distinguishes functions that return as soon as the DMA engine has been programmed (with `_async` suffix) or after all data has been transferred (without suffix). The asynchronous functions allow to start a DMA transfer and then work on different data while the DMA engine completes the transfer. Those functions return a unique transfer identifier, which has to be passed to the `hero_memcpy_wait` function to guarantee transfer completion before the data can be used. Multi-dimensional transfers allow to scatter and gather non-contiguous data with a single function call. For instance, the `hero_memcpy2d_*` functions copy N sequences of B bytes from `src` to `dst` and apply a different address offset to `src` and `dst` after each sequence. This scatter-gather functionality is essential for tiling (e.g., to gather the rows of a tile of a 2D matrix from main memory into a dense SPM buffer before computation and scatter them back after computation). Whenever the DMA engine supports multi-dimensional transfers, they are executed directly by the DMA hardware; otherwise, they are implemented in software.

To measure the performance of applications and their execution on hardware, HEROv2 provides functions that provide a uniform interface to different hardware performance monitors and counters. The functions are mainly designed for hardware counters to which an event is assigned dynamically, which is common in modern processors. The available events range from monotonic clock cycles over memory accesses and stalls to memory and interconnect contention and utilization metrics. The `hero_perf_alloc` function allocates a counter for a given event and resets that counter. If the event is not supported by the hardware or the hardware counters are exhausted, the function returns an error. At the start of a program section to be investigated, a call to `hero_perf_continue_all` starts all allocated counters, and at the end of that section, `hero_perf_pause_all` stops them. Those two functions execute with the minimal latency and overhead supported by the hardware (often as a single inlined CSR write instruction), allowing for precise, fine-grained, and minimally intrusive performance measurements, which are crucial for identifying bottlenecks and systematic optimization.

7.2 Evaluation

Configuration	**Aurora**	**Blizzard**	**Cyclone**
Host ISA	ARMv8.0-A		`RV64GC`
Host Core Arch.	Cortex-A53		CVA6 [ZB19]
Host # Cores	4		1
Accel. ISA	`RV32IMAFCXpulpv2`	`RV32IMAFDXssrXfrepXsdma`	
Accel. Core Arch	CV32E40P [Gau+17]	Snitch [Zar+20]	
Accel. # Cores	8		32
Main Mem. Cap.	4 GiB DDR4		8 GiB HBM2E
Main Mem. BW	up to 19.2 GB/s		up to 460 GB/s
Carrier Silicon	Xilinx ZU9EG		Xilinx VU37P
Carrier Freq.	50 MHz		25 MHz
Status	mature	in development	

Table 7.1: Current target platforms and configurations of HEROv2.

In this section, we evaluate the most mature configuration of HEROv2: As host, it features an industry-standard quad-core 64-bit ARMv8 Cortex-A53 processor with 32 KiB L1 instruction and 32 KiB L1 data cache per core and an 1 MiB L2 cache shared by all four cores, implemented as hard macro and clocked at 1.2 GHz. As PMCA, it features an octa-core 32-bit RISC-V floating-point accelerator (OpenHW CV32E40P core architecture) with 128 KiB L1 SPM and support for custom instructions (`RV32IMAFCXpulpv2`), implemented as soft-macro in the PL. Host and PMCA are connected through a lightweight IOMMU, which allows the PMCA to share the host's virtual memory space and which is implemented as soft-macro in PL, to a shared DRAM controller. The shared main memory consists of 4 GiB DDR4 DRAM, which provides up to 19.2 GB/s of bandwidth.

The implementation of PMCA and IOMMU on the PL of a Xilinx Zynq UltraScale+ ZU9EG SoC achieves a clock frequency of 50 MHz (without any FPGA-specific optimizations). The frequency is mainly limited by paths from the *request* output of the LSU of an accelerator core through the cluster interconnect to the arbitrator of a memory bank and back to *grant* input of the LSU of another core. Among the available PL resources, the configurable logic blocks (CLBs) are the limiting factor with 98.1 % utilization, of which 87.7 % are used by the PMCA and 10.4 % by the IOMMU. Within the PMCA, the cores (each of which includes an FPU), dominate with 38.4 % of the total CLBs. 24.2 % of the block RAM tiles and 2.9 % of the DSP slices are used. We used Xilinx Vivado 2019.2 with the *Alternate Routability* synthesis strategy and the *Congestion–Spread Logic–Low* implementation strategy.

Variants of HEROv2 with alternative host processors and PMCAs are in development, and an overview of current configurations of HEROv2 and their status is shown in Table 7.1. The *Blizzard* configuration shares the host and the carrier silicon with the *Aurora* configuration evaluated here but features an octa-core RISC-V MLT accelerator (`RV32IMAFDXssrXfrepXsdma`) with variable precision support for 8 to 64 bit floating-point numbers. The *Cyclone* configuration targets a larger carrier silicon, on which a multi-cluster configuration of the MLT accelerator fits together with a 64-bit RISC-V host CPU. This configuration will not only offer higher accelerator performance but also an open-source soft-macro host CPU, which contrasts with the "black box" hard-macro Cortex-A53 host CPU of *Aurora* and *Blizzard*.

The evaluated applications and kernels, listed in Table 7.2, represent a wide range of accelerator workloads. From the Polybench/ACC benchmark suite [Gra+12], `2mm`, `3mm`, `atax`, `bicg`, and `gemm` are linear algebra kernels, `conv2d` is part of the "stencil" domain, and `covar` is part of the "datamining" domain. Together, these commonly accelerated kernels span a wide range of memory acess patterns and operational intensities. Additionally, `darknet` is an end-to-end real-time object detection application that implements the YOLO convolutional neural network (CNN) [Red+16]. The data for all applications resides in host-managed shared DRAM. `3mm`, `atax`, `bicg`, and `darknet` (one layer at a time) are composed of consecutive offloads, denoted by arrows ($\rightarrow$) in the table; all other kernels consist of a single offload. All benchmarks are compiled with `-O3` but no specific optimization flags. We take the time stamps of each accelerated application on the host, and it thus includes all data transfers and synchronization between host and accelerator. To measure the accelerator cycles during parts of the application in isolation

Kernel	Accelerated computation	Complexity $\mathcal{O}()$ space	comput.
2mm	$C_{i,j} = \sum_{k=1}^{N} \alpha A_{i,k} B_{k,j}$	N^2 N=1024 ⇒ 12 MiB	N^3
3mm	$E = \text{2mm}(A, B) \rightarrow F = \text{2mm}(C, D) \rightarrow G = \text{2mm}(E, F)$	N^2 N=1024 ⇒ 36 MiB	N^3
atax	$B_i = \sum_{j=1}^{N} A_{i,j} X_j \rightarrow Y_i = \sum_{j=1}^{N} A_{j,i} B_j$	N^2 N=4000 ⇒ 61 MiB	N^2
bicg	$Q_i = \sum_{j=1}^{N} A_{i,j} P_j \rightarrow S_j = \sum_{i=1}^{N} R_i A_{i,j}$	N^2 N=4000 ⇒ 61 MiB	N^2
conv2d	$B_{i,j} = \sum_{(k,l)=(-1,-1)}^{(1,1)} c_{k,l} A_{i+k,j+l}$	N^2 N=2050 ⇒ 16 MiB	N^2
covar	$E_j = \alpha \sum_{i=1}^{M} D_{i,j};\ D_{i,j} \mathrel{-}= E_j;\ S_{i,j} = S_{j,i} = \sum_{k=1}^{N} D_{k,i} D_{k,j}$	N^2 N=1000 ⇒ 4 MiB	N^3
darknet	$C_{i,j} = \sum_{k=1}^{N} \alpha A_{i,k} B_{k,j}$	N^2 34 MiB	N^3
gemm	$C_{i,j} = \beta \left(\sum_{k=1}^{N} \alpha A_{i,k} B_{k,j} \right)$	N^2 N=1024 ⇒ 12 MiB	N^3

Table 7.2: Evaluated kernels and applications. Subscripts denote indices, uppercase letters are variables, and lowercase letters are constants. Arrows (→) denote consecutive offloads. Semicolons (;) denote consecutive computations within the same offload.

(e.g., only computation), we use the hardware performance counter API described in § 7.1.4. In all case studies, the accuracy of all results is fully maintained and verified. In all experiments, the host CPU runs Linux 4.19.0 on a root file system generated with Buildroot 2019.02.1, and we compile applications with LLVM 9.0.0 (extended as described in § 7.1.2).

7.2.1 Application-Level Case Study

We begin with a case study on the application level. For each of the applications introduced above, we want to answer the following questions: How should the local memory of the accelerator be partitioned and data transfers organized so that the run time is dominated by computations on local memory? What is the speed-up compared to letting the accelerator load and store data directly from off-chip main memory? How should the application be parallelized over the cores in the accelerator, and what is the speed-up from parallelization?

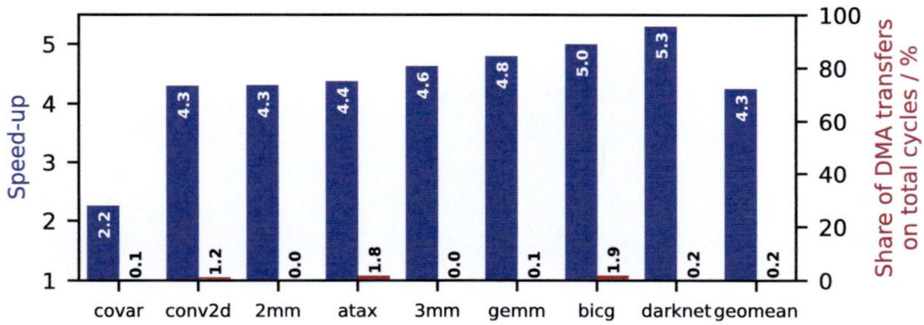

Figure 7.4: Speed-up of execution on local memory with handwritten DMA transfers compared to execution on external main memory. Single accelerator thread.

The first two questions hold the key for making effective use of any accelerator with software-managed local memory. To answer them, we divide input and output data into tiles. Assuming all data have the same dimensionality D, the side length of one tile is given by $S = \lfloor (L/N)^{1/D} \rfloor$, where N is the number of data elements (such as different vectors or matrices) and L is the capacity of the L1 for user data in number of words. With the evaluated accelerator architecture and runtime, $L = 28\,\mathrm{Ki}$ single-precision (i.e., $4\,\mathrm{B}$) words can be stored in L1. Tiling an algorithm is a non-trivial problem to which there is no general solution. We describe the tiling of one algorithm in the following to give an intuition, and we make the source code of all benchmarks available for full transparency and reproducibility (see link in conclusion).

For the convolutional layers in `darknet`, which are implemented as matrix-matrix multiplications, the tile side length of the two input matrices A and B and the output matrix C is $S = 97$. We loop over the tiles of A and transfer the current tile to L1. Within that loop, we loop over the tiles of B corresponding to the current horizontal dimension of A and transfer the corresponding tile of B and C in, perform the tiled matrix-matrix multiplication, and transfer the resulting tile of C out.

The other arithmetic kernels are implemented in an analogous manner. As the left-hand scale in Fig. 7.4 shows, this reduces the run time compared to loading and storing directly from off-chip main memory by 5.3× for `darknet` specifically and by 4.3× on average[4]. While this scheme does not exploit double buffering and the nonblocking DMA transfers that the platform is capable of, the share of cycles spent on DMA transfers is negligible (max: 1.9 %, average: 0.2 %), as the right-hand scale of Fig. 7.4 shows.

Every application lends itself differently to tiling and DMA transfers: In applications with high spatial locality, in particular when computation accesses data in the same sequence as it is stored in memory and it does so for large consecutive arrays, the DMA engine can transfer long continuous data bursts. This is particularly common in linear algebra and CNN kernels: the kernels with the highest speed-up in Fig. 7.4 are all from those domains.

In applications with low spatial locality or divergent access patterns, DMA transfers are substantially shorter and thus offer lower speed-up. Nonetheless, the DMA engine's capability for gather-scatter transfers and many outstanding requests offers a speed-up of more than 4× even with low spatial locality. Temporal locality, on the other hand, has an even bigger impact: For some applications, tiling necessitates that each data element is loaded multiple times because local memory is not large enough to hold all data elements between two use instants. `covar` is an example of such an application, where each element of the data matrix has to be loaded twice (once during mean calculation and once while computing the covariance matrix). This reload factor of two reduces the speed-up by DMA transfers by almost 2× to only 2.2×.

The third question – how an application should be parallelized – holds the key for making effective use of any parallel accelerator. HEROv2's OpenMP runtime library enables to answer this question efficiently by experimentation: For the computation on one tile, we simply annotate the outermost computational loop with `#pragma omp for` to distribute its execution over the cores of the accelerator. As the left bar for each application in Fig. 7.5 shows, this reduces the computation cycles by 6.5 to 7.1× (average: 6.9×) on an 8-core cluster.

[4]Whenever we discuss the *average* of normalized numbers, we mean the *geometric mean* (denoted *geomean* in the figures), as reasoned in [FW86].

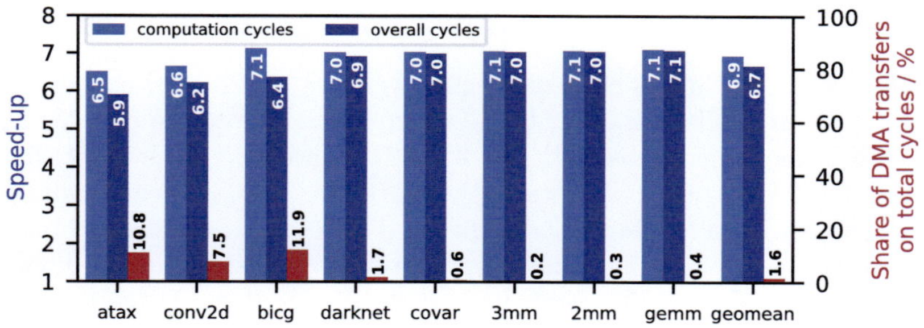

Figure 7.5: Speed-up of execution with 8 accelerator threads compared to execution with 1 accelerator thread. Execution on local memory with handwritten DMA transfers.

Even higher speed-ups by parallelization could be achieved by optimizing the loop schedule and stride, but we set that aside because it is sensitive to data size. The overall application speed-up by parallelization, shown by the middle bar for each application, is between 5.9 to 7.1× (average: 6.7×). The right bar shows why the computation-only speed-up cannot be achieved for the overall application: The DMA transfers are not sped up by parallelization, so their share on the total cycles increases by the overall speed-up factor. Due to Amdahl's law, this limits the overall speed-up achievable by parallelization. On average, 1.6 % of cycles spent on DMA transfers result in a modest decrease from 6.9× to 6.7×. However, for some applications, such as `bicg`, 11.9 % of cycles spent on DMA transfers reduce the parallelization speed-up from 7.1× to 6.4×. This may justify a more complex double-buffered implementation of an application.

This benchmark analysis shows how HEROv2's full-stack hardware and software allows to rapidly explore and optimize the accelerated performance of domain-relevant applications on a heterogeneous computer prototype: The high emulation throughput allows to study realistic problem sizes, and the complete software stack allows to adapt and tune real-world applications and representative kernels with reasonable effort and make informed optimization decisions. Furthermore, the fully open hardware implementation allows tracing and profiling hardware, as well as optimizing it.

7.2.2 Runtime-Level and Toolchain Case Study

Tiling an algorithm for efficient execution on accelerator-local memory is not only an intellectual effort but also requires extra code to be written, verified, and maintained. HEROv2's API is designed to simplify this task for device-specific operations such as DMA transfers, which can be executed with a single function call, and fork-join parallelism, which is available through the standardized OpenMP pragmas that call into the runtime library. However, the part of tiling that is specific to each algorithm cannot be substantially simplified by a runtime library.

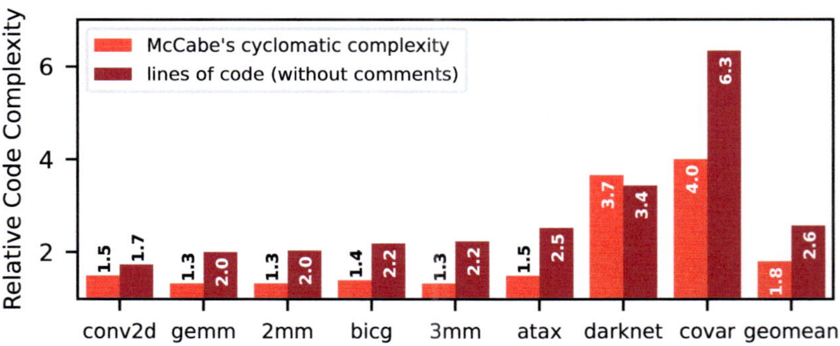

Figure 7.6: Code complexity with handwritten tiling and DMA transfers compared to the unmodified code of each application. The light red bars show McCabe's cyclomatic complexity and the dark red bars show lines of code without comments.

The code complexity increase by handwritten tiling and DMA transfers compared to unmodified code is shown in Fig. 7.6 for each application. We used the CCCC tool[5] [Lit01] on the accelerated part of each application and extracted two of its results: (1) The lines of code (without comments), which can be an indication for the effort of writing and reading a piece of code. (2) McCabe's cyclomatic complexity, which counts the number of linearly independent paths through a piece of code, and which can be an indication for the effort of understanding and verifying a piece of code. The results show three coarse categories of applications: First, the six applications on the left are tiled in a

[5] `https://sourceforge.net/projects/cccc`

single dimension, which is a modest effort: the lines of code increase by 1.7 to 2.5× and the cyclomatic complexity increases by 1.3 to 1.5×. On average, the lines of code overhead by 1D tiling is 2× and the cyclomatic complexity increase is 1.4×. Second, `darknet` with its CNN layers is implemented with two-dimensional tiling and DMA transfers. 2D tiling substantially increases both the cyclomatic complexity (3.7×) and the lines of code (3.4×). Third, `covar` is also implemented with 2D tiling, but the implementation is additionally split over two separate iterations through the entire data. This means the ca. 3× lines of code overhead by 2D tiling incurs twice, leading to a total 6.3× lines of code overhead, while the cyclomatic complexity increases by the same factor as for `darknet`. In summary, the additional effort and maintenance cost for tiling an algorithm ranges from modest (1.7× LOC, 1.5× cyclo. compl.) to very high (6.3× LOC, 4.0× cyclo. compl.) and is certainly not negligible on average (2.6× LOC, 1.8× cyclo. compl.).

OpenMP assumes a cache-based memory hierarchy, leading to low performance on SPM-based memory hierarchies if a program is not manually tiled. To save these substantial manual tiling efforts, an optimal solution would be if the toolchain could automatically transform the untiled algorithm code to manage the memory hierarchy. The *AutoDMA* feature, introduced in § 7.1.2, brings this to HEROv2. Effectively, this means that the software-managed memory hierarchy of HEROv2 can be programmed as easily as a cache-based system.

The speed-up of compiler-generated and handwritten tiled code over unmodified OpenMP code is shown in Fig. 7.7. While the handwritten tiled code has a significantly higher complexity than the unmodified OpenMP code, as shown in Fig. 7.6, compiler-generated tiling requires zero code changes. The benchmarks in Fig. 7.7 can be divided into two categories: For `covar` and `atax`, the speed-up achieved by the compiler is marginal. For all other benchmarks, the speed-up achieved by the compiler is comparable to that of handwritten code. The benchmarks in the latter category feature large segments of contiguous memory accesses (spatial locality), and achieve on average 85 % of the speed-up of handwritten code. The remaining 15 % come from leveraging programmer insights (i.e., information not expressed in the code) to reduce the number of reconfigurations of the DMA engine: The handwritten code transfers multiple rows of matrices at once, possible by the understanding that the first element of row $N + 1$ is next in

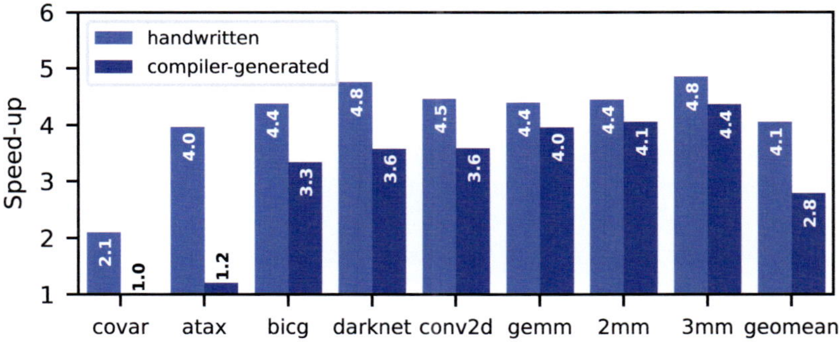

Figure 7.7: Speed-up of execution on local memory with *compiler-generated* tiling and DMA transfers compared to execution on external main memory. 8 accelerator threads. The light blue bars show the speed-up of the handwritten implementation for comparison; the dark blue bars show the speed-up achieved by the compiler.

memory to the last element of row N. The compiler was not able to reconstruct this information, due to *array-to-pointer decay* in which the dimensions of data structures are lost. Without this information, the compiler considers multiple rows as non-contiguous and initiates a new DMA burst for each row, which adds an overhead compared to the single DMA burst in the handwritten code. Nonetheless, AutoDMA provides a speed-up of up to 4.4× without any code changes. Expert programmers still have the option to turn this feature off and implement tiling manually to extract the last tens of percent of performance.

For two benchmarks (`covar` and `atax`), the compiler-generated code cannot compete with the handwritten code. This can also be attributed to memory access patterns: a significant part of memory accesses are performed column-wise, i.e., in non-contiguous blocks. This effect is aggravated by the tile shape selected by the compiler, which inadvertently maximizes the number of column-wise accesses per tile, rather than contiguous row-wise accesses. This is due to the loop ordering of the benchmarks, which the AutoDMA feature does not rewrite[6]. Spatial

[6]Tools that reorder loops, such as polyhedral analyses and transformations [GGL12], could be used to preprocess the code, or the benchmarks could be manually rewritten using classical spatial locality optimizations.

locality is also important for performance on cache-based systems, but the issue is aggravated on HEROv2 where the DMA engine in this case is used to transfer individual words. As such, it is an extreme case of the overhead discussed for the previous category, where the compiler could not find sufficiently large chunks of contiguous memory. Despite these problems, the performance with AutoDMA is on-par to up to 20 % higher than the OpenMP baseline, due to the high bandwidth of the DMA engine.

In summary, for programs with high spatial locality, HEROv2's AutoDMA feature provides performance comparable to handwritten code, without the need for explicit tiling and DMA transfers. This reduces the execution time of unmodified OpenMP programs by up to 4.4× on software-managed memory hierarchies, achieving 85 % of the speed-up of handwritten code. This makes software-manged memory hierarchies as easy to program as their hardware-cache-based counterparts. Similarly to hardware-managed caches, AutoDMA provides no significant improvements for programs with low spatial locality.

HEROv2 is a unique platform to analyze, develop, and optimize such compiler and runtime techniques, because it allows executing real applications and reference benchmarks on the actual RTL logic of a heterogeneous SoC, and because all its hardware and software components are open-source and permissively licensed.

7.2.3 System Architecture-Level Case Study

Our third case study examines the impact of an architectural design decision: How does the data width of the accelerator into the shared interconnect and main memory influence the performance of accelerated applications? To answer this question, we customize the on-chip network of the accelerator once to half the data width (32 bit) and once to twice the data width (128 bit) and remeasure our applications.

Fig. 7.8 shows the speed-up (for values > 1) or slow-down (for values < 1) for an accelerator on-chip network data width of 32 bit (left three bars of each application) and 128 bit (right three bars) compared to 64 bit. The leftmost bar in each group of three bars compares the cycles spent on DMA transfers: For most applications, halving the data width of the on-chip network results in a speed-up of 0.5×, and doubling the data width results in a speed-up of 2×, as expected. The exception,

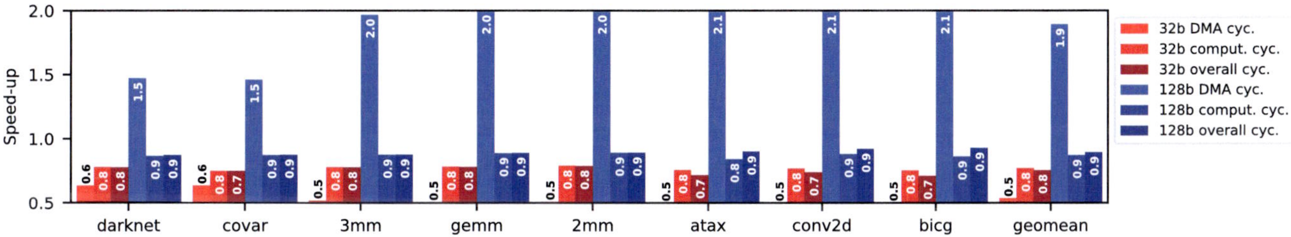

Figure 7.8: Speed-up of execution with an accelerator on-chip network data width of 32 bit or 128 bit compared to the default of 64 bit. Each application has six bars: The three bars on the left are for 64 bit data width, the three bars on the right are for 128 bit data width. Of each three bars, the left shows the speed-up of only the DMA cycles, the middle of only the computation cycles, and the right of the total cycles.

however, is `darknet`, with 0.6× for half the data width and 1.5× for double the data width. `covar` and `darknet` are the only applications to use two-dimensional DMA transfers, which are composed of many relatively short bursts. This transfer pattern does not fully saturate the given on-chip network, which results in lower speed-ups for wider data widths. That is an important insight for optimizing the on-chip network if DMA performance was critial for application performance. However, as we know from the application-level case study (§ 7.2.1), DMA transfers only account for at most 11.9 % (average: 1.6 %) of the application cycles. The majority of cycles is spent in computations, and the middle of each three bars compares the cycles spent on computations: Surprisingly, the data width of the on-chip network also has a significant impact on them. For 32 bit, the fetch bandwidth of instructions into the cache is halved, which leads to more instruction fetch stall cycles and reduces computational performance. For 128 bit, the fetch bandwidth for instructions could be doubled, but the instruction cache can only fetch at most 64 bit per cycle, so that has no impact. To accommodate the wider memory interface of the DMA engine, the tightly-coupled data memory (TCDM) interconnect in the accelerator cluster has to be changed from 14×16 to 18×32. This configuration causes on average 14 % more contention on the TCDM despite the higher number of banks. A careful realignment of the cores on the TCDM interconnect could alleviate this, but the gist is that a wider accelerator on-chip network does not automatically increase performance. In fact, as the rightmost bar of each application shows, application performance decreases by 10 % on average if the design of the cluster is not simultaneously adapted.

Such insights from fully measured application executions are central for making substantiated decisions on the system architecture and for prioritizing engineering efforts. The closer the measured prototype is to the final design, the higher the quality of the measurements. Effects such as those discussed in this section would be extremely difficult to model with a simulator, as they depend on fine-grained interaction between several hardware components. Capturing this interaction quantitatively with non-cycle-accurate architectural simulation is a very intricate and error-prone task. Thus, an application-programmable heterogeneous research platform with a complete hardware and software stack, such as HEROv2, is a key enabler for architecture-level performance exploration.

7.2.4 Accelerator ISA-Level Case Study

Specialized instructions are an important part of many domain-specific accelerators. They are often designed and evaluated in an instruction set simulation (ISS) or in RTL simulations. The drawback of ISS is that it is inaccurate as performance model because it does not capture microarchitectural effects. RTL simulation models the microarchitecture accurately, but it is only feasible for small data set and does not take communication outside the accelerator, which influences the memory subsystem and thereby the execution of the accelerated kernel, into account. Thus, a heterogeneous research platform is required to quantify the impact of specialized accelerator ISA extensions in heterogeneous computing with real-world data sets.

In this case study, we answer the question "How much do instructions from the `Xpulpv2` ISA extension speed up execution of heterogeneous applications compared to the standard `rv32imafc` ISA?". As described in § 7.1.2, we have extended the RISC-V LLVM backend to automatically emit `Xpulpv2` instructions during machine code generation. The evaluated kernels process data at full precision (i.e., 32-bit integers or floats) and therefore cannot make use of the quarter- or half-precision packed SIMD instructions, which would offer a significant speed-up for reduced-precision processing.

The speed-up of the `Xpulpv2` ISA extension over the standard RISC-V `RV32IMAFC` ISA is shown in Fig. 7.9. We measure the total accelerator cycles with handwritten DMA transfers and 8 accelerator threads. As the first bar of each application shows, simply enabling `Xpulpv2` provides a speed-up of 1.5× on average. Starting with `gemm` as an example, we find that the compiler replaces the inner two compute loops by hardware loops. This is optimal, as there is only hardware for two loops. The body of the innermost loop is halved from 10 instructions (2 loads, 4 additions, 2 multiplications, 1 store, and 1 branch) to 5 instructions (2 post-increment loads, 1 multiplication, 1 MAC, and 1 store), while the bodies of the outer levels stay mostly identical. Apart from the store, which could be hoisted out of the innermost loop by a memory-to-register optimization pass, the innermost loop is optimal[7], and it is also optimally scheduled. The resulting speed-up of 2.5×

[7]For `gemm`, the multiplication by α could be hoisted out of the innermost loop for all data types where multiplication is distributive over addition. However, this is an

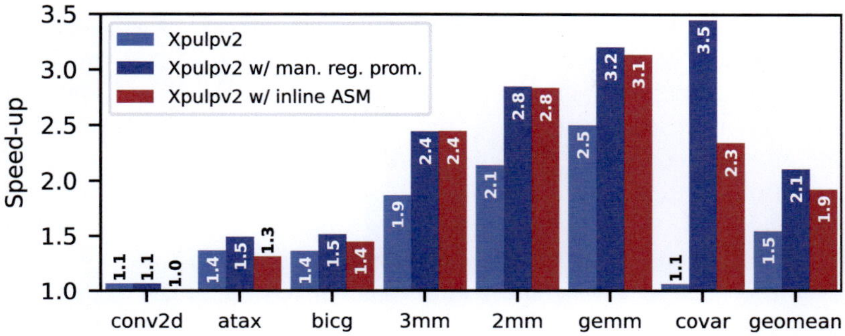

Figure 7.9: Speed-up of execution with custom instructions (Xpulpv2) compared to the standard RISC-V (RV32IMAFC) ISA. Execution on local memory with handwritten DMA transfers and 8 accelerator threads. For each application, the first bar shows the speed-up by Xpulpv2 instructions without manual register promotion, the second bar by Xpulpv2 instructions with manual register promotion, and the third bar by implementing the innermost loop with inline assembly instructions (including manual register promotion and Xpulpv2 instructions).

can be attributed to halving the instructions within the innermost loop (ca. 2× speed-up) and hardware loops as well as less instructions in the outer loops (ca. 0.5× speed-up). Manually hoisting the store out of the innermost loop significantly improves performance further. Again looking at `gemm`, this reduces the innermost loop from 5 to 4 instructions, and the resulting relative speed-up of 1.28× is aligned with the reduction in instructions. The same findings hold for `3mm` and `2mm`, and a comparison with an inline assembly implementation of the innermost loop reveals that the instructions generated by the compiler perform on-par or better than the expert-written instructions. However, some benchmarks behave quite differently: For `conv2d`, `atax`, and `bicg`, the Xpulpv2 ISA extension provides only between 10 to 50 % of speed-up – both with compiler-generated instructions and with an expert-written inner loop body. There are two main reasons for this: First, the kernels are not as well suited for post-increment memory

algebraic transformation and does not apply to all data types (such as `floats`), so we do not consider it.

accesses as the matrix-matrix multiplication kernels. For `atax`, the increment of one of the two loads in the innermost loop is too large to be used in post-increment. For `conv2d`, the 2D (3×3) loads in the innermost loop leave some opportunities for post-increment loads, but they are complex to exploit. Even the expert-written instructions, which use as many post-increment accesses as possible, do not bring a significant speed-up. Second, hardware loops are not inferred for the innermost loops. This could be because the innermost loop iterates over the rows in a tile, and the number of rows changes depending on the tile index. This does not fundamentally preclude the use of hardware loops, however, so it is a current compiler limitation. Finally, `covar` sees a very high speed-up with `Xpulpv2`, but only with manual memory-to-register promotion. This simple change in the code enables the compiler to infer a hardware loop. The instructions generated by the compiler substantially outperform the expert-written inner loop, due to better scheduling.

In summary, the `Xpulpv2` ISA extension has the potential to significantly accelerate all kernels we evaluated, mainly through post-increment memory accesses and hardware loops. Especially the latter are not trivial for the compiler to generate in all cases, however, which currently leads to speed-ups between 1.1 to 3.5× (average: 2.1×). While the impact of changes to the accelerator ISA could also be studied in isolation (e.g., in RTL simulation), evaluating within a heterogeneous prototype system (such as HEROv2) produces more representative results, because the balance and interaction between memory transfers from and to the shared main memory and computation are taken into account, and because a prototype running at tens of MHz makes it feasible to work with real-scale data sets.

7.3 Related Work

Emulation systems on FPGAs or custom programmable logic are widely used to get cycle-accurate results at a clock frequency of multiple MHz and turn-around times of few hours to days. In the FAME taxonomy [Tan+10a], HEROv2 is a *Direct FAME* system, which are characterized by implementing the target system with a one-to-one correspondence in clock cycles on an FPGA. Commercial Direct FAME

Platform	FAME	Emul. Focus	HW Ecosys.	Open-Source	Heterogeneous SoC Features: Hosts	Accelerators	Program. Model	Accel. Coh.	MDM
Cadence Palladium [Cad21]	0	Dig	©	✗	customer-provided				
Siemens Veloce Strato [Sie21]	0	Dig	©	✗	customer-provided				
Synopsys ZeBu [Syn21]	0	Dig	©	✗	customer-provided				
OpenPiton [Bal+16; Bal+20a; Bal+20b]	0	Many	p	✓	T1, CVA6, PRV32, ao486	MIAOW, NVDLA	[OS driver]	4	✗
MEG [Zha+20]	0	Mem	p	✓	Boom	custom	n/d	n/d	✗
DART [Wan+14; PPT21], DuCNoC [KAH18]	3	Net	d	✗	n/a			n/d	✗
FireSim [Kar+18; Kar+20]	1, 5	Dig	[Chipyard]	✓	Rocket	NVDLA, HLS	[OS driver]	n/d	✗
Centrifuge [Hua+19]	1, 5	HeSoC	[Chipyard]	✓	Rocket	HLS	[OS driver]	2, 4	✗
OpenESP [Man+20; Gir+21]	0	HeSoC	p	✓	CVA6, LEON3	custom, HLS	[user-space API]	1, 2, 4	✗
HEROv1 [Kur+17; Kur+18c]	0	HeSoC	[PULP], [OpenHW]	✓	A9	[PULP]	[OS driver], [user-space API], [unified]	1, 2	✗
HEROv2 [this chapter]	0	HeSoC	[PULP], [OpenHW]	✓	A53, CVA6	[PULP], [Snitch]	[OS driver], [user-space API], [unified]	1, 2, 3	✓

Table 7.3: Comparison of computer emulation systems on programmable logic devices. Legend: **FAME**: taxonomy numbers [Tan+10a] | **Emul. Focus**: Digital hardware, Heterogeneous SoCs, Manycores, Near-Memory Processing, On-Chip Networks | **HW Ecosys.**: © commercial, as ported or developed by platform maintainers, [PULP] PULP, [OpenHW] OpenHW Group, [Chipyard] Chipyard | **Hosts**: ARM Cortex-A9 or -A53, OpenHW Group CVA6, OpenSPARC T1, Berkeley out-of-order machine (Boom), Gaisler LEON3, PicoRV32 | **Accelerators**: [PULP] PULP cluster, [Snitch] Snitch cluster, MIAOW GPGPU, custom logic, NVDLA, HLS-generated | **Programming Model**: accelerator separately programmable through [OS driver] an OS driver or [user-space API] a user-space API, or [unified] unified host+accelerator programming with heterogeneous OpenMP applications, or n/d not defined | **Accelerator Coherence Modes** [Zuc+21]: 1 non-coherent DMA, 2 LLC-coherent DMA, 3 coherent DMA, 4 coherent cache, or not defined (n/d) | **MDM**: mixed-data-model (e.g., 64-bit host + 32-bit accelerator) programming supported.

systems include Cadence Palladium [Cad21], Siemens Veloce Strato [Sie21], and Synopsys ZeBu [Syn21]. Those systems are capable of emulating up to 20 billion ASIC gate equivalents (GEs) at up to 10 MHz and can cost millions of USD. HEROv2 can scale over multiple FPGAs with chip-to-chip FPGA Mezzanine Card (FMC) connections, which are supported by all of HEROv2's carrier silicon. Depending on the design, the system interconnect, the accelerator interconnect, or both can extend over multiple FPGAs through FMC and QSFP+ connections. HEROv2's currently largest carrier silicon, Xilinx' VCU128, offers ca. 40 million ASIC GEs and can communicate with other FPGAs at more than 650 Gbit/s. Depending on FPGA and configuration, HEROv2's clock frequency is between 20 to 100 MHz.

FPGA-based computer system emulators are common in industry and research. The following recent works are comparable with ours (see Table 7.3 for an overview and [Ang+14] for a broader survey of older approaches up to 2014): OpenPiton [Bal+16] is an open-source many-core research framework that can be implemented on an FPGA. It comes with a cache-coherent on-chip network and by now supports four different processor cores [Bal+20b; Bal+20a], among them CVA6 also supported by HEROv2. The most recent version of OpenPiton optionally includes an open-source GPGPU or Nvidia's deep learning accelerator (NVDLA), which can be programmed from Linux running on the processor cores. This recent developments allow using OpenPiton for research on heterogeneous computing, which is HEROv2's focus, but the full hardware-software stack integration of accelerators, from API to accelerator-specific compiler backend, remains HEROv2's distinguishing feature. MEG [Zha+20] is a system emulation infrastructure for near-data processing implemented on an FPGA. It features four 64-bit RISC-V Boom cores as host processor and a near-memory accelerator whose architecture and ISA are not specified. Like HEROv2, MEG features a Linux-booting host processor and is implemented on a VU37P, but unlike HEROv2, the focus is on near-memory accelerators that seem to have a fixed function, as accelerator programming, memory hierarchy, data transfers, and communication with the host are not discussed. DART [Wan+14] accelerates the simulation of on-chip networks by mapping them onto an FPGA. It provides programmability by decoupling the simulator architecture from the architecture of the simulated on-chip network. Similarly, DuCNoC [KAH18] maps on-chip networks to the PL of a Zynq-7000 SoC. Like HEROv2, the on-chip network is highly configurable and modeled cycle-accurately at 10 MHz and more in DART and DuCNoC, but unlike HEROv2, the remainder of the computer system remains in a higher-level simulator that injects traffic into the on-chip network. Prasad *et. al.* [PPT21] improve on DART by specializing the microarchitecture of on-chip network components to the target FPGA architecture, which reduces the required hardware resources by 70 % and the average packet latency by 20 %. In contrast, HEROv2's components are not specialized to FPGAs, which means they consume more hardware resources than minimally required but also that they match an ASIC implementation cycle-by-cycle.

FireSim [Kar+18; Kar+20] extends FPGA-based emulation to Amazon EC2 F1, a public cloud FPGA platform. On the FPGA of each instance, FireSim allows instantiating modules from the Chipyard [Ami+20] ecosystem (e.g., the 64-bit RISC-V Rocket core, a L2 cache, a NIC, and fixed-function accelerators such as the Hwacha vector processor). Multiple instances are connected over the datacenter network and C++ simulation models to emulate datacenter clusters with multiple server nodes. Like HEROv2, FireSim comes with an OS-capable multi-core CPU, but unlike HEROv2, the focus is on datacenter clusters and networking instead of heterogeneous computing with different ISAs, data models, execution models, and memory subsystems. Centrifuge [Hua+19] extends FireSim with a flow that generates heterogeneous SoCs containing user-defined high-level synthesis (HLS) accelerators together with a Linux driver for them. In contrast, accelerators in HEROv2 can be interfaced with user-space libraries or in heterogeneous OpenMP applications, but the accelerator software is not auto-generated.

Research platforms that combine HW and SW components are less common. OpenESP [Man+20; Gir+21] is a research platform for heterogeneous SoC design. It provides a methodology and components to integrate processors (among them CVA6 also supported by HEROv2) and HLS-generated accelerators with a 2D-mesh on-chip network. Like in HEROv2, the accelerators have a DMA engine and can share virtual addresses with a processor through an IOMMU and a Linux driver. Unlike in HEROv2, accelerators are not programmable with a full-featured standard ISA, and there is thus no OpenMP offloading support and no heterogeneous API, runtime libraries, and toolchain that span across host processors and accelerators. HEROv1 [Kur+17; Kur+18c] does provide the components that enable the evaluation of heterogeneous applications on a mixed-ISA computer, but its toolchain is fundamentally limited to 32-bit hosts and accelerators [Kur+20b]. Additionally, it has no API that unifies programming over multiple accelerators; it features one host and one accelerator architecture, and hardware and software are tailored to those instead of being modular; and its on-chip network is limited to simple configurations (e.g., fixed 64-bit data width) and topologies (e.g., central crossbar), which do not meet the demands of modern heterogeneous computers.

Accelerators have been designed specifically for FPGAs. GRVI Phalanx [Gra16] is a 32-bit RISC-V soft processor array that scales to more than 1000 cores on a Xilinx VU9P FPGA. 2GRVI Phalanx [Gra19] extends that to more than 1000 64-bit RISC-V cores on a Xilinx VU37P. The DRAGON architecture [AYB21] is a 64-bit custom-ISA cluster-based multiprocessor that scales to 144 cores on a Xilinx VU37P. In contrast, the accelerator in HEROv2 is not specialized for FPGAs but has identical RTL code as for ASIC tapeouts. Its components, from cores [Gau+17; ZB19] over the accelerator cluster [Ros+17] to the on-chip communication fabric [Kur+22] have been taped out in multiple ASICs. An open-source GPGPU, such as MIAOW [Bal+15], could be added as an accelerator to HEROv2. Enabling full support for OpenMP offloading, as is supported by LLVM for CUDA-based GPUs, would require developing an LLVM backend and offloading plugin for that GPU. In the HPC domain, IBM's A2 core, which powers the Blue Gene/Q supercomputers, has been released open-source as OpenPOWER A2I/A2O cores[8]. These cores could be used in HEROv2 as host or accelerator. Linux and LLVM support the Power ISA, so extending HEROv2 to support them would require integrating the A2 core into the hardware system (compliant with its memory interface) and software stack (e.g., writing an accelerator driver and runtime library).

Programming models targeting heterogeneous computing are manifold, and we refer to [MV15] for an overview. In OpenCL [Khr21b], an application on the host submits separately-written kernels to be executed on an accelerator to a command queue. OpenCL is imperative, meaning application programmers have to explicitly call functions to create buffers, transfer data, and start execution on an accelerator.

---

[8] `https://github.com/openpower-cores/a2i`

SYCL [Khr21a] extends OpenCL by enabling single-source heterogeneous programming and C++ AMP [Mic18] by relieving the programmers from explicit data transfers between host and device. oneAPI Data Parallel C++ (DPC++) [Int20b] builds on SYCL to define functions that can be offloaded to devices, and an open-source LLVM implementation is in development. OpenMP, supported natively by HEROv2, is declarative, meaning application programmers describe *what* they want to do (e.g., offload a code section with data to an accelerator) while the compiler and runtime libraries take care of *how* those actions happen. OpenACC [OpenACC3.1] goes even further: its directives describe the properties of a program (e.g., a parallel loop with independent iterations), and the toolchain and runtime libraries specialize the program to an accelerator. In Clang, OpenACC is implemented by translation to OpenMP. Through this, HEROv2 also supports OpenACC. Fortran, which is relevant in HPC [Loh10], could in the future be supported through LLVM's Flang project, which aims to support offloading by the end of 2022 [Cha21]. HEROv2's open-source LLVM-based toolchain will enable the community to construct complementary and alternative heterogeneous computing software stacks, while building on a solid open infrastructure.

Heterogeneous compilers have also been implemented by others. Intel offers an OpenMP offloading compiler for its Xeon Phi accelerators [Int18], which differ from the host CPU by accelerator-specific extensions. Those extensions are only available through the proprietary Intel compiler, whereas HEROv2's full toolchain is open source. Research works on GCC [CMB18] were the first to provide an open-source heterogeneous OpenMP toolchain, but GCC's offloading compilation is fundamentally limited to the same data model (e.g., 32-bit) for host and accelerators [Kur+20b]. Mixed-data-model heterogeneous compilation has been pioneered recently [Kur+20b] with Clang/LLVM, and HEROv2 integrates that work into its toolchain.

7.4 Summary

HEROv2 is a full-stack open-source[9] research platform for state-of-the-art heterogeneous computing: HEROv2 provides all hardware and

[9] https://github.com/pulp-platform/hero

software required to develop, compile, and run single-source, single-binary heterogeneous applications and seamlessly offload and share data from an application-class 64-bit host to a programmable 32-bit parallel accelerator. Thus, HEROv2 enables effective and accurate research from applications and algorithms down to microarchitecture. Additionally, HEROv2 comes with a novel *AutoDMA* compiler plugin, which provides a solution to one of the most pressing problems of accelerators with software-managed memories: without any code changes, AutoDMA tiles loops and infers DMA transfers, which leads to a speed-up of up to 4.4× without any code changes and in most cases is only 15 % slower than a handwritten implementation, which requires 2.6× more code.

HEROv2 enables research in various domains, and we know of ongoing projects that use HEROv2 in high-performance computing, real-time processing, in-network processing, transprecision accelerators, and parallel programming. We expect future work to evolve in the directions of larger scale-out accelerators, mixed and finer-grained coherency domains, and novel virtualization and communication technologies. We are also working on a tape-out in a modern silicon technology.

Chapter 8

Conclusions

This thesis presented the first open-source research platform for HeSoCs. Our heterogeneous research platform, HERO, integrates all essential components of a HeSoC – from 32-bit accelerator cores capable of sharing a 64-bit virtual address space with the host, over on-chip network IPs and heterogeneous compiler toolchains, to operating system support and a unified application programming interface. HERO thereby enables revisiting the entire computing stack, which is seen as a prerequisite to solve the many open challenges keeping us from exploiting the full potential of heterogeneous computing [Zah17].

This thesis additionally advances the state of the art and public knowledge on four main components of HeSoCs: on-chip communication, atomic memory operations, virtual memory sharing, and heterogeneous compilation and offloading. The remainder of this chapter summarizes the main results of this thesis and identifies unsolved challenges and opportunities for future work.

8.1 Main Results

On-Chip Communication

On-chip communication infrastructure is a central component of modern SoCs, and it continues to gain importance as the number of cores, the heterogeneity of components, and the on-chip and off-chip bandwidth continue to grow. Decades of research on on-chip networks enabled cache-coherent shared-memory multiprocessors. However, communication fabrics that meet the needs of heterogeneous many-core and accelerator-rich SoCs, which are not (or only partially) coherent, are a much less mature research area.

To fill this knowledge gap, we presented our platform for high-performance on-chip communication, which enables the construction of heterogeneous many-core and accelerator-rich SoCs independent of closed-source on-chip network IPs. Our work advances the state of the art through two main contributions: First, network (de)multiplexers as elementary components make the design and verification of custom network modules substantially easier. Second, an end-to-end palette of modules from a DMA engine to on-chip memory controllers, including data and ID width converters, as well as the widest range of data widths and concurrent transactions, enables new designs.

We characterized the absolute and the asymptotic complexity of all of our network modules for the most important parameters (such as data width, number of IDs, number of ports, and number of concurrent transactions). The critical path of all modules scales at worst linearly in their parameters, for most modules and parameters even only logarithmically. As the absolute results of the minimum clock period show, the critical path of all modules remains below 500 ps post-topographical-synthesis in a modern 22 nm FDSOI technology in the large design space that we evaluated. This shows our modules are suited for a wide range of target frequencies and bandwidths, up to 2 to 3 GHz. The area of most modules scales linearly in their parameters, with the notable exception of the ID width, which causes an exponential growth of the demultiplexer and all modules containing it. As the absolute results show, most modules fit within a few tens of kGE when not pushed to the highest possible clock frequency and parametrization. Even more complex modules, such as a 4×4 crossbar with up to 256 independent concurrent transactions, fit in a modest 100 kGE when clocked at 2.5 GHz. Finally, we used our platform to design and implement a state-of-the-art 1024-core MLT accelerator in a 22 nm FDSOI technology, where our communication fabric provides 32 TB/s cross-sectional bandwidth at only 24 ns round-trip latency between any two cores.

Atomic Memory Operations

Atomic memory operations (AMOs) are ubiquitous in modern concurrent algorithms, but their scalable implementation is not a solved problem: First, their implementation in commercial processors is a well-guarded secret; thus there is a knowledge gap on the challenges and trade-offs of implementing AMOs. Second, the subsystem for executing AMOs is presumably tightly coupled to the processor architecture and the memory hierarchy. Finally, AMOs on modern multiprocessors have been shown to scale poorly to large numbers of threads [SBH15; ELF11].

To resolve those limitations and fill the knowledge gap, we presented our ATomic UNit (ATUN), an open-source hardware module, to implement AMOs at any level in the memory hierarchy. Our work advances the state of the art through two main contributions: First, the ATUN gives system designers and even application and library programmers full control over where in a software-managed memory hierarchy AMOs should be resolved. This enables keeping shared data at its original location while PEs send AMOs to the memory holding the data, where the AMOs are executed locally. Second, the ATUN decouples the execution of AMOs and conditional-store-based primitives from locking shared resources as much as possible. This allows our solution to scale the throughput of AMOs linearly until the target memory saturates.

We evaluated our ATUN on a cycle-accurate FPGA prototype, where 32 hardware threads (harts) share a second-level SPM, and determined the scalability and performance of individual operations and entire concurrent algorithms. First, the throughput of AMOs scales linearly with the number of harts until the on-chip memory is saturated. This holds with and without contention. Second, the latency of an AMO is only 25 % higher than a regular load from that memory. Under contention, the latency increases linearly with 10 cycles per concurrently accessing core. Third, the throughput of important concurrent algorithms scales linearly with the number of harts until the memory bandwidth is saturated. We synthesized our ATUN for a 22 nm FDSOI technology for a variable number of harts in the system and find that its area increases linearly at only 0.5 kGE per core and its longest path scales logarithmically with the number of harts. The longest path through our ATUN does not increase with the number of harts. A few paths grow logarithmically with the number of harts, but they do not become critical for the evaluated maximum of 128 harts. This makes the our ATUN well suited to scale to a large number of harts at a very low hardware cost per hart. This is a major step towards solving the problem of scalable AMOs in software-managed memory hierarchies.

Shared Virtual Memory

Shared virtual memory (SVM) is key in HeSoCs, both for programmability and to avoid data duplication. However, SVM can incur a significant execution time overhead when TLB entries are missing. Moreover, allowing DMA burst transfers to write SVM traditionally requires hardware buffers to absorb transfers that miss in the TLB. These buffers have to be overprovisioned for the maximum burst size, wasting precious on-chip memory, and stall all SVM accesses once they are full, hence hampering the scalability of parallel accelerators.

To resolve those limitations, we presented our SVM solution that avoids the majority of TLB misses with prefetching, supports parallel burst DMA transfers without additional buffers, and can be scaled with the workload and number of parallel processors. Our solution is based on three novel concepts: First, to minimize the rate of TLB misses, the TLB is proactively filled by compiler-generated prefetch helper threads, which use run-time information to issue timely prefetches. Second, to reduce the latency of TLB misses, misses are handled by a variable number of parallel miss handling threads. Third, to support parallel burst DMA transfers to SVM without additional buffers, we add lightweight hardware to a standard DMA engine (less than 10 % extra module area) to react to TLB misses.

Our evaluation shows that, compared to the prior state of the art [Vog+17], our work improves the PMCA performance by up to 4× for irregular memory access patterns and by up to 60 % for regular memory access patterns. Compared to using hardware buffers to absorb bursts from DMA engines in a conventional IOMMU, our solution requires two orders of magnitude less memory and scales better, as it only stalls the DMA engine that caused a TLB miss.

Heterogeneous Compilation and Offloading

The optimal address width of host and PMCAs continues to diverge: While the host manages ever more application memory, PMCAs are designed to work mainly on their local memory. Today, 64-bit hosts are common, but only few PMCAs (especially among those integrated in HeSoCs) exceed 32-bit addressable local memory, a difference expected to increase with 128-bit hosts in the exascale era. Managing this discrepancy requires support for multiple *data models* in heterogeneous compilers. So far, compiler support for multiple data models has not been explored, which hinders the programmability of mixed-data-model HeSoCs and inhibits their adoption.

To resolve this limitation, we designed and implemented the first mixed-data-model compiler, supporting arbitrary address widths on host and PMCA. To hide the inherent complexity and to enable high programmer productivity, our LLVM-based compiler supports transparent offloading on top of OpenMP. This includes four main contributions: First, we showed that unless GCC fundamentally changes its approach to heterogeneous compilation and representing pointers of different widths in its IR, mixed-data-model compilation is infeasible in GCC. Second, we showed that LLVM's separation of address spaces with different data layouts enables mixed-data-model compilation. Third, we presented and implemented algorithms to assign pointers to the native device address space. Fourth, we designed and implemented a legalization pass that allows a 32-bit core to access 64-bit addresses with minimal hardware support and without any programmer intervention.

Our evaluation on a 64+32-bit HeSoC shows that memory can be transparently shared between host and PMCA at overheads below 0.7 % compared to 32-bit-only execution. Since the overhead primarily occurs for setup sequences, where a PMCA must access 64-bit addresses, it is amortized and becomes negligible for real-scale data sets, where local memory is primarily accessed. Our solution thus enables mixed-data-model computers to execute at near-native performance.

Open-Source Heterogeneous Research Platform

Realizing the full potential of heterogeneous computers still requires solving many challenges. A research platform that serves as a prototype is an effective way to work on these challenges, but none existed for the current generation of heterogeneous computers prior to this thesis.

To fill this gap, we presented HERO, our FPGA-based research platform that combines PMCAs composed of 32-bit RISC-V cores with application-class 64-bit host processors. HERO allows to seamlessly share data between host and PMCA and comes with an open-source high-performance on-chip network, a unified heterogeneous programming interface, and a mixed-data-model and mixed-ISA heterogeneous compiler based on LLVM. HERO additionally includes a novel solution to one of the most pressing problems in heterogeneous computing: how to relieve the programmer of the burden of specializing an algorithm to the memory hierarchy of the PMCA.

We evaluated HERO in four case studies from the application level over toolchain and system architecture down to PMCA microarchitecture, and we showed how HERO enables effective research and development on the full stack of heterogeneous computing. For instance, the compiler, through a novel *AutoDMA* pass, can tile loops and infer data transfers to and from the PMCA, which leads to a speed-up of up to 4.4× compared to the original program and in most cases is only 15 % slower than a handwritten implementation, which requires 2.6× more lines of code.

8.2 Outlook

This thesis focused on the creation of an open-source research platform for HeSoCs, along with the design and implementation of central HeSoC components that were not described in literature before. We expect that this lays the foundation for a lot of research on heterogeneous computing in the future. In fact, that work has only just begun. The following are a few leads to guide potential future developments.

Bring HERO to Silicon. A central limitation of the FPGA-based HERO is that it cannot directly be used to measure power and energy consumption. This is because the circuit on the FPGA is equivalent to that on a custom-made IC in terms of cycle-by-cycle behavior but not in terms of logic cells and their connections. Bringing HERO to silicon in a custom-made IC would allow these measurements. Additionally, there is a gap in maximum clock frequency between components implemented as hard macro and those implemented on an FPGA. Having both host and PMCAs as hard macros in an IC allows direct comparisons without having to compensate for the clock frequency difference.

Explore Chiplet-Based Heterogeneous Computing. Chiplets are widely seen as a solution for increasing performance despite the slowing of Moore's Law scaling [Moo19]. Besides providing more transistors, chiplets additionally have the potential to increase the heterogeneity of system-in-a-package (SiP) computers: different PMCAs from multiple vendors, each on its own chiplet, could be integrated together with various kinds of memory technology chiplets in a single package. This creates new design and optimization opportunities and challenges. To tackle them, HERO could be extended over multiple FPGAs and to chiplet carrier boards with high-speed transceivers.

Achieve Performance Portability. Getting an algorithm to perform well on a heterogeneous computer still requires a lot of manual work and knowledge on each combination of algorithm and computer. Incorporating most of that work and knowledge into automated processes would ideally enable each algorithm to exploit the full potential of every computer. HERO can be used for developments on compilers [Lat+21a], programming models [Omp5.1], and higher-level frameworks [Ben+19] to reach that goal.

Find the Boundaries of Heterogeneity. The boundaries of heterogeneity in computing are still mostly uncharted territory. For instance, how does the principle of diminishing marginal utility apply to accelerator nesting? For example, does it make sense to add a fixed-function discrete wavelet transform (DWT) unit to a cluster in a PMCA that is dedicated to image signal processing, or should more PEs be added to the cluster instead? Using optimization frameworks [Mor+14], agile hardware development methodologies [Lee+16], compiler-based hardware-software co-design [Lat+21b], or all of them together with HERO to rapidly evaluate realistic prototypes could bring a breakthrough on these questions.

Appendix A

Acronyms

ABI application binary interface.

ACE AXI Coherency Extensions.

ACP Accelerator Coherency Port.

AES Advanced Encryption Standard.

AI artificial intelligence.

ALU arithmetic logic unit.

AMO atomic memory operation.

API application programming interface.

AS address space.

ASIC application-specific integrated circuit.

AST abstract syntax tree.

AT area and timing.

ATUN ATomic UNit.

AXI Advanced eXtensible Interface.

BRAM block random access memory.

CDC clock domain crossing.

CHI Coherent Hub Interface.

CLB configurable logic block.

CNN convolutional neural network.

CPU central processing unit.

CSR control and status register.

CU compute unit.

D2D die-to-die link.

DDG data dependency graph.

DDR double data rate.

DMA direct memory access.

DRAM dynamic random-access memory.

DRC design rule checking.

DSO dynamic shared object.

DSP digital signal processor.

DWC data width converter.

DWT discrete wavelet transform.

ELF executable and linkable format.

FAME FPGA Architecture Model Execution.

FDSOI fully depleted silicon on insulator.

FF flip-flop.

FIFO first-in first-out buffer.

FLOP floating-point operation.

FMC FPGA Mezzanine Card.

FPGA field-programmable gate array.

FPU floating-point unit.

FSM finite state machine.

GE gate equivalent.

GF22FDX GlobalFoundries' 22 nm fully-depleted silicon-on-insulator.

GPGPU general-purpose graphics processing unit.

GPU graphics processing unit.

HAL hardware abstraction library.

hart hardware thread.

HBM High Bandwidth Memory.

HDL hardware description language.

HeSoC heterogeneous system on chip.

HLS high-level synthesis.

HPC high-performance computing.

HSA Heterogeneous System Architecture.

I/O input/output.

IC integrated circuit.

IOMMU input/output memory management unit.

IoT internet of things.

IP intellectual property module.

IPA interprocedural analysis.

IR intermediate representation.

ISA instruction set architecture.

ISS instruction set simulation.

Juno ADP Juno ARM Development Platform.

LDS linked data structure.

LLC last level cache.

LR load-reserved.

LSU load/store unit.

LTO link-time optimization.

LUT lookup table.

LZC leading-zero counter.

MAC multiply-accumulate operation.

MHT miss handling thread.

ML machine learning.

MLT machine learning training.

MMU memory management unit.

MSB most significant bit.

NIC network interface controller.

NN neural network.

NoC network on chip.

OCCP on-chip communication protocol.

OS operating system.

PCIe Peripheral Component Interconnect Express.

PE processing element.

PHT Prefetching Helper Thread.

PHY physical layer.

PL programmable logic.

PMCA programmable many-core accelerator.

PTW page table walker.

QoS quality of service.

RAB Remapping Address Block.

RF register file.

RMW read-modify-write.

RTE runtime environment.

RTL register-transfer level.

RVTSO RISC-V Total Store Order.

RVWMO RISC-V Weak Memory Order.

SC store-conditional.

SIMD single instruction multiple data.

SIMT single instruction multiple threads.

SiP system-in-a-package.

SMMU system memory management unit.

SoA state of the art.

SoC system on chip.

SOI silicon on insulator.

SPM scratchpad memory.

SPMD single program multiple data.

SRAM static random access memory.

SSA static single assignment.

SVM shared virtual memory.

TCDM tightly-coupled data memory.

TLB translation lookaside buffer.

TSO total store ordering.

VLIW very long instruction word.

VMM virtual memory management.

VPU vision processing unit.

WT Worker Thread.

Appendix B

Bibliography

[ABC03] H. Al-Sukhni, I. Bratt, and D. A. Connors. "Compiler-Directed Content-Aware Prefetching for Dynamic Data Structures." In: *IEEE PACT '03*. PACT '03. Washington, DC, USA: IEEE Computer Society, 2003, pp. 91–. ISBN: 0-7695-2021-9. URL: http://dl.acm.org/citation.cfm?id=942806.943843.

[ABW20] S. Arora, D. Bouvier, and C. Weaver. "AMD Next Generation 7nm Ryzen 4000 APU "Renoir"." In: *2020 IEEE Hot Chips 32 Symposium (HCS)*. Los Alamitos, CA, USA: IEEE Computer Society, Aug. 2020, pp. 1–30. DOI: 10.1109/HCS49909.2020.9220414. URL: https://doi.ieeecomputersociety.org/10.1109/HCS49909.2020.9220414.

[Acc13] Accellera Inc. *Open Core Protocol Specification Release 3.0.* 2013.

[Adi14] A. Adinets. *CUDA Pro Tip: Optimized Filtering with Warp-Aggregated Atomics.* 2014.

[AMD18] AMD Corp. *AMD Radeon Instinct MI60.* Datasheet. Nov. 2018. URL: https://www.amd.com/system/files/documents/radeon-instinct-mi60-datasheet.pdf.

[AMD20a] AMD Inc. *AMD CDNA Architecture: The All-New AMD GPU Architecture for the Modern Era of HPC & AI.* White Paper. 2020. URL: https://www.amd.com/system/files/documents/amd-cdna-whitepaper.pdf.

[AMD20b] AMD Inc. *AMD High-Performance Embedded GPUs: The Optimal Balance of Processing Performance, Power Efficiency, and Cost Effectiveness.* Product Brief. 2020. URL: https://www.amd.com/system/files/documents/high-performance-gpu-product-brief.pdf.

[Amd67] Gene M. Amdahl. "Validity of the Single Processor Approach to Achieving Large Scale Computing Capabilities." In: *Proceedings of the April 18-20, 1967, Spring Joint Computer Conference.* AFIPS '67 (Spring). Atlantic City, New Jersey: Association for Computing Machinery, 1967, pp. 483–485. ISBN: 9781450378956. DOI: 10.1145/1465482.1465560. URL: https://doi.org/10.1145/1465482.1465560.

[Ami+20] Alon Amid et al. "Chipyard: Integrated Design, Simulation, and Implementation Framework for Custom SoCs." In: *IEEE Micro* 40.4 (2020), pp. 10–21. DOI: 10.1109/MM.2020.2996616.

[Ang+14] Hari Angepat, Derek Chiou, Eric S. Chung, and James C. Hoe. "FPGA-Accelerated Simulation of Computer Systems." In: *Synthesis Lectures on Computer Architecture* 9.2 (2014), pp. 1–80. DOI: 10.2200/S00586ED1V01Y201407CAC029. URL: https://doi.org/10.2200/S00586ED1V01Y201407CAC029.

[Ant+16] Samuel F. Antao et al. "Offloading Support for OpenMP in Clang and LLVM." In: *LLVM-HPC'16.* Salt Lake City, Utah, 2016. ISBN: 978-1-5090-3878-7. DOI: 10.1109/LLVM-HPC.2016.6.

[APJ09] Niket Agarwal, Li-Shiuan Peh, and Niraj K. Jha. "In-Network Coherence Filtering: Snoopy Coherence without Broadcasts." In: *IEEE/ACM MICRO.* 2009. ISBN: 9781605587981. DOI: 10.1145/1669112.1669143.

[App+07] R. W. Apperson, Z. Yu, M. J. Meeuwsen, T. Mohsenin, and B. M. Baas. "A Scalable Dual-Clock FIFO for Data Transfers Between Arbitrary and Haltable Clock Domains." In: *IEEE TVLSI* (2007).

[ARM16] ARM Ltd. *ARM CoreLink MMU-500 System Memory Management Unit.* Technical reference manual. Cambridge, UK, Feb. 2016.

[Arm17a] Arm Ltd. *AMBA AXI and ACE Protocol Specification.* 2017.

[Arm17b] Arm Ltd. *Arm Mali Midgard OpenCL Developer Guide, version 3.6.* Dec. 2017. URL: https://developer.arm.com/documentation/100614/0306/.

[Arm19a] Arm Ltd. In: *Architecture Reference Manual: ARMv8 for ARMv8-A architecture profile.* Arm Ltd., July 2019. Chap. D1.19 Interprocessing.

[Arm19b] Arm Ltd. *ARMv8-A Architecture Reference Manual.* 2019.

[AS19] A. Akram and L. Sawalha. "A Survey of Computer Architecture Simulation Techniques and Tools." In: *IEEE Access* 7 (2019), pp. 78120–78145. DOI: 10.1109/ACCESS.2019.2917698.

[Asa+16] Krste Asanović et al. *The Rocket Chip Generator.* Tech. rep. EECS Department, University of California, Berkeley, Apr. 2016.

[AXI-F.b] *AMBA AXI and ACE Protocol Specification Issue F.b.* Arm Ltd. 2017.

[AYB21] Riadh Ben Abdelhamid, Yoshiki Yamaguchi, and Taisuke Boku. "A Highly-Efficient and Tightly-Connected Many-Core Overlay Architecture." In: *IEEE Access* 9 (2021), pp. 65277–65292. DOI: 10.1109/ACCESS.2021.3074171.

[Bal+15] Raghuraman Balasubramanian, Vinay Gangadhar, Ziliang Guo, Chen-Han Ho, Cherin Joseph, Jaikrishnan Menon, Mario Paulo Drumond, Robin Paul, Sharath Prasad, Pradip Valathol, and Karthikeyan Sankaralingam. "MIAOW – An open source RTL implementation of a GPGPU." In: *2015 IEEE Symposium in Low-Power and High-Speed Chips (COOL CHIPS XVIII)*. 2015, pp. 1–3. DOI: 10.1109/CoolChips.2015.7158663.

[Bal+16] Jonathan Balkind, Michael McKeown, Yaosheng Fu, Tri Nguyen, Yanqi Zhou, Alexey Lavrov, Mohammad Shahrad, Adi Fuchs, Samuel Payne, Xiaohua Liang, Matthew Matl, and David Wentzlaff. "OpenPiton: An Open Source Manycore Research Framework." In: *ACM ASPLOS*. Atlanta, Georgia, USA, 2016. ISBN: 9781450340915. DOI: 10.1145/2872362.2872414.

[Bal+20a] J. Balkind, T. Chang, P. J. Jackson, G. Tziantzioulis, A. Li, F. Gao, A. Lavrov, G. Chirkov, J. Tu, M. Shahrad, and D. Wentzlaff. "OpenPiton at 5: A Nexus for Open and Agile Hardware Design." In: *IEEE Micro* 40.4 (2020), pp. 22–31. DOI: 10.1109/MM.2020.2997706.

[Bal+20b] Jonathan Balkind et al. "BYOC: A "Bring Your Own Core" Framework for Heterogeneous-ISA Research." In: *Proceedings of the Twenty-Fifth International Conference on Architectural Support for Programming Languages and Operating Systems*. ASPLOS '20. Lausanne, Switzerland: Association for Computing Machinery, 2020, pp. 699–714. ISBN: 9781450371025. DOI: 10.1145/3373376.3378479. URL: https://doi.org/10.1145/3373376.3378479.

[Ban+02] R. Banakar, S. Steinke, Bo-Sik Lee, M. Balakrishnan, and P. Marwedel. "Scratchpad memory: a design alternative for cache on-chip memory in embedded systems." In: *Proceedings of the Tenth International Symposium on Hardware/Software Codesign. CODES*

2002 (IEEE Cat. No.02TH8627). 2002, pp. 73–78. DOI: 10.1145/774789.774805.

[Ban+19] Pete Bannon, Ganesh Venkataramanan, Debjit Das Sarma, Emil Talpes, and Bill McGee. “Computer and Redundancy Solution for the Full Self-Driving Computer.” In: *2019 IEEE Hot Chips 31 Symposium (HCS)*. Los Alamitos, CA, USA: IEEE Computer Society, Aug. 2019, pp. 1–22. DOI: 10.1109/HOTCHIPS.2019.8875645. URL: https://doi.ieeecomputersociety.org/10.1109/HOTCHIPS.2019.8875645.

[BD06] Luca Benini and Giovanni De Micheli. *Networks on Chips: Technology and Tools*. Morgan Kaufmann, 2006. ISBN: 978-0-12-370521-1.

[Bea+02] Olivier Beaumont, Vincent Boudet, Fabrice Rastello, and Yves Robert. “Partitioning a Square into Rectangles: NP-Completeness and Approximation Algorithms.” In: *Algorithmica* (June 2002). DOI: 10.1007/s00453-002-0962-9.

[Bea+19] Olivier Beaumont, Brett A. Becker, Ashley DeFlumere, Lionel Eyraud-Dubois, Thomas Lambert, and Alexey Lastovetsky. “Recent Advances in Matrix Partitioning for Parallel Computing on Heterogeneous Platforms.” In: *IEEE Transactions on Parallel and Distributed Systems* 30.1 (2019), pp. 218–229. DOI: 10.1109/TPDS.2018.2853151.

[Ben+19] Tal Ben-Nun, Johannes de Fine Licht, Alexandros N. Ziogas, Timo Schneider, and Torsten Hoefler. “Stateful Dataflow Multigraphs: A Data-Centric Model for Performance Portability on Heterogeneous Architectures.” In: *Proceedings of the International Conference for High Performance Computing, Networking, Storage and Analysis*. SC ’19. Denver, Colorado: Association for Computing Machinery, 2019. ISBN: 9781450362290. DOI: 10.1145/3295500.3356173. URL: https://doi.org/10.1145/3295500.3356173.

[Ber+17] Gheorghe-Teodor Bercea, Carlo Bertolli, Arpith C. Jacob, Alexandre Eichenberger, Alexey Bataev, Georgios Rokos, Hyojin Sung, Tong Chen, and Kevin O'Brien. "Implementing Implicit OpenMP Data Sharing on GPUs." In: *Proceedings of the Fourth Workshop on the LLVM Compiler Infrastructure in HPC*. LLVM-HPC'17. Denver, CO, USA: ACM, 2017, 5:1–5:12. ISBN: 978-1-4503-5565-0. DOI: 10.1145/3148173.3148189. URL: http://doi.acm.org/10.1145/3148173.3148189.

[Bin+11] Nathan Binkert et al. "The Gem5 Simulator." In: *SIGARCH Comput. Archit. News* 39.2 (Aug. 2011), pp. 1–7. ISSN: 0163-5964. DOI: 10.1145/2024716.2024718. URL: https://doi.org/10.1145/2024716.2024718.

[Bly20] D. Blythe. "The Xe GPU Architecture." In: *2020 IEEE Hot Chips 32 Symposium (HCS)*. Los Alamitos, CA, USA: IEEE Computer Society, Aug. 2020, pp. 1–27. DOI: 10.1109/HCS49909.2020.9220591. URL: https://doi.ieeecomputersociety.org/10.1109/HCS49909.2020.9220591.

[Boh+17] B. Bohnenstiehl et al. "KiloCore: A 32-nm 1000-Processor Computational Array." In: *IEEE Journal of Solid-State Circuits (JSSC)* 52.4 (Apr. 2017), pp. 891–902.

[Bor+13] Daniele Bortolotti, Christian Pinto, Andrea Marongiu, Martino Ruggiero, and Luca Benini. "VirtualSoC: A Full-System Simulation Environment for Massively Parallel Heterogeneous System-on-Chip." In: *2013 IEEE International Symposium on Parallel & Distributed Processing, Workshops and Phd Forum*. IEEE. 2013, pp. 2182–2187.

[Bra+14] A. Bradbury et al. *Tagged Memory and Minion Cores in the lowRISC SoC*. Dec. 2014. URL: http://www.lowrisc.org/downloads/lowRISC-memo-2014-001.pdf.

[Bur+19] T. Burd, N. Beck, S. White, M. Paraschou, N. Kalyanasundharam, G. Donley, A. Smith, L. Hewitt, and S. Naffziger. "Zeppelin: An SoC for Multichip Architectures." In: *IEEE JSSC* (2019).

[But+16] Anastasiia Butko, Florent Bruguier, Abdoulaye Gamatié, Gilles Sassatelli, David Novo, Lionel Torres, and Michel Robert. "Full-System Simulation of big.LITTLE Multicore Architecture for Performance and Energy Exploration." In: *2016 IEEE 10th International Symposium on Embedded Multicore/Many-core Systems-on-Chip (MCSOC)*. 2016, pp. 201–208. DOI: 10.1109/MCSoC.2016.20.

[But+18] Anastasiia Butko, Albert Chen, David Donofrio, Farzad Fatollahi-Fard, and John Shalf. "Open2C: Open-Source Generator for Exploration of Coherent Cache Memory Subsystems." In: *ACM MEMSYS*. 2018. ISBN: 9781450364751. DOI: 10.1145/3240302.3270314.

[BV20] L. Bajic and J. Vasiljevic. "Compute substrate for Software 2.0." In: *2020 IEEE Hot Chips 32 Symposium (HCS)*. Los Alamitos, CA, USA: IEEE Computer Society, Aug. 2020, pp. 1–31. DOI: 10.1109/HCS49909.2020.9220687. URL: https://doi.ieeecomputersociety.org/10.1109/HCS49909.2020.9220687.

[Cad18] Cadence Design Systems, Inc. *Tensilica Xtensa LX7 processor datasheet.* Feb. 2018. URL: https://ip.cadence.com/uploads/1099/TIP_PB_Xtensa_lx7_FINAL-pdf.

[Cad21] *Palladium Z2 Enterprise Emulation Platform – Enabling design teams to debug and accelerate verification of complex IPs to multi-billion-gate SoC designs.* Cadence. 2021. URL: https://www.cadence.com/en_US/home/tools/system-design-and-verification/emulation-and-prototyping/palladium.html#z2.

[Cav+20] Matheus Cavalcante, Andreas Kurth, Fabian Schuiki, and Luca Benini. "Design of an Open-Source Bridge between Non-Coherent Burst-Based and Coherent Cache-Line-Based Memory Systems." In: *Proceedings of the 17th ACM International Conference on Computing Frontiers*. CF '20. Catania, Sicily, Italy: Association for Computing Machinery, 2020, pp. 81–88. ISBN: 9781450379564. DOI: 10.1145/3387902.3392631. URL: https://doi.org/10.1145/3387902.3392631.

[Cel+15] C. Celio et al. *The Berkeley Out-of-Order Machine (BOOM): An Industry-Competitive, Synthesizable, Parameterized RISC-V Processor*. June 2015. URL: http://www2.eecs.berkeley.edu/Pubs/TechRpts/2015/EECS-2015-167.html.

[Cel18] Christopher Celio. "A Highly Productive Implementation of an Out-of-Order Processor Generator." en. PhD thesis. Berkeley, CA, USA: University of California, Berkeley, 2018.

[CES71] E. G. Coffman, M. Elphick, and A. Shoshani. "System Deadlocks." In: *ACM Comp. Surv.* (1971). ISSN: 0360-0300. DOI: 10.1145/356586.356588.

[CF16] Alessandro Cilardo and Edoardo Fusella. "Design automation for application-specific on-chip interconnects: A survey." In: *Integration* (2016). ISSN: 0167-9260. DOI: https://doi.org/10.1016/j.vlsi.2015.07.017.

[CG20] J. Choquette and W. Gandhi. "NVIDIA A100 GPU: Performance & Innovation for GPU Computing." In: *2020 IEEE Hot Chips 32 Symposium (HCS)*. Los Alamitos, CA, USA: IEEE Computer Society, Aug. 2020, pp. 1–43. DOI: 10.1109/HCS49909.2020.9220622. URL: https://doi.ieeecomputersociety.org/10.1109/HCS49909.2020.9220622.

[Cha21] K. Chandramohan. "LLVM/Flang: OpenMP update." In: *Sixth OpenMP Developers Conference (OpenMP Con)*. Bristol, UK, Sept. 2021. URL: https://openmpcon.org/wp-content/uploads/openmpcon2021-flang.pdf.

[Che+15] Tianshi Chen, Zidong Du, Ninghui Sun, Jia Wang, Chengyong Wu, Yunji Chen, and Olivier Temam. "A High-Throughput Neural Network Accelerator." In: *IEEE Micro* 35.3 (2015), pp. 24–32. DOI: 10.1109/MM.2015.41.

[CHI-D] *AMBA5 CHI Specification Issue D.* Arm Ltd. 2019.

[Chi+12] N. Chitlur et al. "QuickIA: Exploring Heterogeneous Architectures on Real Prototypes." In: *IEEE HPCA*. Feb. 2012, pp. 1–8. DOI: 10.1109/HPCA.2012.6169046.

[Chi15] David Chisnall. "Adventures with LLVM in a magical land where pointers are not integers." In: *2015 LLVM Developer's Meeting*. Feb. 2015. URL: https://llvm.org/devmtg/2015-02/slides/chisnall-pointers-not-int.pdf.

[Cho+16] Young-kyu Choi, Jason Cong, Zhenman Fang, Yuchen Hao, Glenn Reinman, and Peng Wei. "A Quantitative Analysis on Microarchitectures of Modern CPU-FPGA Platforms." In: *Proceedings of the 53rd Annual Design Automation Conference*. DAC '16. Austin, Texas: ACM, 2016, 109:1–109:6. ISBN: 978-1-4503-4236-0. DOI: 10.1145/2897937.2897972. URL: http://doi.acm.org/10.1145/2897937.2897972.

[CJ17] S. Chandrasekaran and G. Juckeland. *OpenACC for Programmers: Concepts and Strategies*. Pearson Education, 2017. ISBN: 9780134694344.

[CJG02] R. Cooksey, S. Jourdan, and D. Grunwald. "A Stateless, Content-directed Data Prefetching Mechanism." In: *Proceedings of the 10th International Conference on Architectural Support for Programming Languages and Operating Systems*. ASPLOS X. San

Jose, California: ACM, 2002, pp. 279–290. ISBN: 1-58113-574-2. DOI: 10.1145/605397.605427. URL: http://doi.acm.org/10.1145/605397.605427.

[CM17] A. Capotondi and A. Marongiu. "Enabling Zero-copy OpenMP Offloading on the PULP Many-core Accelerator." In: *Proceedings of the 20th International Workshop on Software and Compilers for Embedded Systems*. SCOPES '17. Sankt Goar, Germany: ACM, 2017, pp. 68–71. ISBN: 978-1-4503-5039-6. DOI: 10.1145/3078659.3079071. URL: http://doi.acm.org/10.1145/3078659.3079071.

[CMB18] Alessandro Capotondi, Andrea Marongiu, and Luca Benini. "Runtime Support for Multiple Offload-Based Programming Models on Clustered Manycore Accelerators." In: *IEEE Transactions on Emerging Topics in Computing* 6.3 (2018), pp. 330–342. DOI: 10.1109/TETC.2016.2554318.

[Cod13] L. Codrescu. "Qualcomm Hexagon DSP: An architecture optimized for mobile multimedia and communications." In: *2013 IEEE Hot Chips 25 Symposium (HCS)*. Los Alamitos, CA, USA: IEEE Computer Society, Aug. 2013, pp. 1–23. DOI: 10.1109/HOTCHIPS.2013.7478317. URL: https://doi.ieeecomputersociety.org/10.1109/HOTCHIPS.2013.7478317.

[Cod15] Lucian Codrescu. "Architecture of the Hexagon 680 DSP for Mobile Imaging and Computer Vision." In: *2015 IEEE International Symposium on High Performance Chips (HOTCHIPS '27)*. 2015. URL: https://www.hotchips.org/wp-content/uploads/hc_archives/hc27/HC27.24-Monday-Epub/HC27.24.20-Multimedia-Epub/HC27.24.211-Hexagon680-Codrescu-Qualcomm.pdf.

[Col+02] J. Collins, S. Sair, B. Calder, and D. M. Tullsen. "Pointer Cache Assisted Prefetching." In: *Proceedings of the 35th Annual ACM/IEEE International Symposium on Microarchitecture*. MICRO 35. Istanbul, Turkey: IEEE Computer Society Press, 2002, pp. 62–73. ISBN: 0-7695-1859-1. URL: http://dl.acm.org/citation.cfm?id=774861.774869.

[Coo12] S. Cook. *CUDA Programming: A Developer's Guide to Parallel Computing with GPUs*. Applications of GPU Computing Series. Elsevier Science, 2012. ISBN: 9780124159884.

[Cot+15] E. G. Cota, P. Mantovani, G. Di Guglielmo, and L. P. Carloni. "An analysis of accelerator coupling in heterogeneous architectures." In: *2015 52nd ACM/EDAC/IEEE Design Automation Conference (DAC)*. June 2015, pp. 1–6. DOI: 10.1145/2744769.2744794.

[CSB18] F. Conti, P. D. Schiavone, and L. Benini. "XNOR Neural Engine: A Hardware Accelerator IP for 21.6-fJ/op Binary Neural Network Inference." In: *IEEE Transactions on Computer-Aided Design of Integrated Circuits and Systems* (2018). ISSN: 0278-0070. DOI: 10.1109/TCAD.2018.2857019.

[CSS11] Shuai Che, Jeremy W. Sheaffer, and Kevin Skadron. "Dymaxion: Optimizing Memory Access Patterns for Heterogeneous Systems." In: *Proceedings of 2011 International Conference for High Performance Computing, Networking, Storage and Analysis*. SC '11. Seattle, Washington: Association for Computing Machinery, 2011. ISBN: 9781450307710. DOI: 10.1145/2063384.2063401. URL: https://doi.org/10.1145/2063384.2063401.

[Cum08] Clifford E Cummings. "Clock domain crossing (CDC) design & verification techniques using SystemVerilog." In: *SNUG*. 2008.

[Cut16] Ian Cutress. "CEVA Launches Fifth-Generation Machine Learning Image and Vision DSP Solution: CEVA-XM6." In: *AnandTech* (Sept. 2016). URL: https://www.anandtech.com/show/10700.

[CV20] Zhaowei Cai and Nuno Vasconcelos. "Rethinking Differentiable Search for Mixed-Precision Neural Networks." In: *Proceedings of the IEEE/CVF Conference on Computer Vision and Pattern Recognition (CVPR).* June 2020.

[CW15] Stephen P. Crago and John Paul Walters. "Heterogeneous Cloud Computing: The Way Forward." In: *Computer* 48.1 (2015), pp. 59–61. DOI: 10.1109/MC.2015.14.

[Dam+15] Marvin Damschen, Heinrich Riebler, Gavin Vaz, and Christian Plessl. "Transparent Offloading of Computational Hotspots from Binary Code to Xeon Phi." In: *Proc. of the 2015 Design, Automation & Test in Europe Conf. & Exh.* DATE '15. Grenoble, France: EDA Consortium, 2015, pp. 1078–1083. URL: http://dl.acm.org/citation.cfm?id=2757012.2757063.

[Dav+18] S. Davidson et al. "The Celerity Open-Source 511-Core RISC-V Tiered Accelerator Fabric: Fast Architectures and Design Methodologies for Fast Chips." In: *IEEE Micro* (2018).

[Day+14] B. K. Daya, C. O. Chen, S. Subramanian, W. Kwon, S. Park, T. Krishna, J. Holt, A. P. Chandrakasan, and L. Peh. "SCORPIO: A 36-core research chip demonstrating snoopy coherence on a scalable mesh NoC with in-network ordering." In: *ACM/IEEE ISCA*. 2014.

[De +02] G. De Micheli, R. Ernst, W. Wolf, and M. Wolf. *Readings in Hardware/Software Co-Design.* Electronics & Electrical. Elsevier Science, 2002. ISBN: 9781558607026.

[De +19] Simon Garcia De Gonzalo, Sitao Huang, Juan Gomez-Luna, Simon Hammond, Onur Mutlu, and Wen-mei Hwu. "Automatic generation of warp-level primitives and atomic instructions for fast and portable parallel reduction on GPUs." In: *CGO*. 2019.

[Di +19] Salvatore Di Girolamo, Konstantin Taranov, Andreas Kurth, Michael Schaffner, Timo Schneider, Jakub Beránek, Maciej Besta, Luca Benini, Duncan Roweth, and Torsten Hoefler. "Network-Accelerated Non-Contiguous Memory Transfers." In: *Proceedings of the International Conference for High Performance Computing, Networking, Storage and Analysis*. SC '19. Denver, Colorado: Association for Computing Machinery, 2019. ISBN: 9781450362290. DOI: 10.1145/3295500.3356189. URL: https://doi.org/10.1145/3295500.3356189.

[Di +21] Salvatore Di Girolamo, Andreas Kurth, Alexandru Calotoiu, Thomas Benz, Timo Schneider, Jakub Beranek, Luca Benini, and Torsten Hoefler. "A RISC-V in-network accelerator for flexible high-performance low-power packet processing." In: *Proceedings of the 48th IEEE/ACM Annual International Symposium on Computer Architecture*. ISCA '21. New York, NY, USA: ACM, June 2021.

[Din+13] B. D. de Dinechin et al. "A Clustered Manycore Processor Architecture for Embedded and Accelerated Applications." In: *IEEE HPEC*. Sept. 2013, pp. 1–6. DOI: 10.1109/HPEC.2013.6670342.

[DKR18] Michael Ditty, Ashish Karandikar, and David Reed. "Nvidia's Xavier SoC." In: *2018 IEEE International Symposium on High Performance Chips (HOTCHIPS '30)*. 2018. URL: https://www.hotchips.org/hc30/1conf/1.12_Nvidia_XavierHotchips2018Final_814.pdf.

[DLV18] Joel E Denny, Seyong Lee, and Jeffrey S Vetter. "Clacc: Translating OpenACC to OpenMP in Clang." In: *2018 IEEE/ACM 5th Workshop on the LLVM Compiler Infrastructure in HPC (LLVM-HPC)*. IEEE. 2018, pp. 18–29.

[DM98] L. Dagum and R. Menon. "OpenMP: an industry standard API for shared-memory programming." In:

IEEE Computational Science and Engineering 5.1 (1998), pp. 46–55. DOI: 10.1109/99.660313.

[DT03] William Dally and Brian Towles. *Principles and Practices of Interconnection Networks.* Morgan Kaufmann, 2003. ISBN: 9780080497808.

[DTH20] William J. Dally, Yatish Turakhia, and Song Han. "Domain-Specific Hardware Accelerators." In: *Commun. ACM* 63.7 (July 2020), pp. 48–57. ISSN: 0001-0782. DOI: 10.1145/3361682. URL: https://doi.org/10.1145/3361682.

[Dur+11] Alejandro Duran, Eduard Ayguadé, Rosa M. Badia, Jesús Labarta, Luis Martinell, Xavier Martorell, and Judit Planas. "OmpSs: A Proposal for Programming Heterogeneous Multi-Core Architectures." In: *Parallel Processing Letters* 21.02 (2011), pp. 173–193. DOI: 10.1142/S0129626411000151. URL: https://doi.org/10.1142/S0129626411000151.

[ELF11] M. Elteir, H. Lin, and W. Feng. "Performance Characterization and Optimization of Atomic Operations on AMD GPUs." In: *2011 IEEE Cluster.* 2011.

[EMP09] E. Ebrahimi, O. Mutlu, and Y. N. Patt. "Techniques for bandwidth-efficient prefetching of linked data structures in hybrid prefetching systems." In: *2009 IEEE 15th International Symposium on High Performance Computer Architecture.* Feb. 2009, pp. 7–17. DOI: 10.1109/HPCA.2009.4798232.

[EPL08] N. D. Enright Jerger, L. Peh, and M. H. Lipasti. "Virtual tree coherence: Leveraging regions and in-network multicast trees for scalable cache coherence." In: *IEEE/ACM MICRO.* 2008.

[EPS06] N. Eisley, L. Peh, and L. Shang. "In-Network Cache Coherence." In: *IEEE/ACM MICRO.* 2006.

[Fat+16] F. Fatollahi-Fard, D. Donofrio, G. Michelogiannakis, and J. Shalf. "OpenSoC Fabric: On-chip network generator." In: *IEEE ISPASS.* 2016.

[FB10] J. Flich and D. Bertozzi. *Designing Network On-Chip Architectures in the Nanoscale Era.* CRC Press, 2010. ISBN: 9781439837115.

[FBM18] Björn Forsberg, Luca Benini, and Andrea Marongiu. "HePREM: Enabling predictable GPU execution on heterogeneous SoC." In: *2018 Design, Automation & Test in Europe Conference & Exhibition (DATE).* 2018, pp. 539–544. DOI: 10.23919/DATE.2018.8342066.

[FBM20] Björn Forsberg, Luca Benini, and Andrea Marongiu. "HePREM: A Predictable Execution Model for GPU-based Heterogeneous SoCs." In: *IEEE Transactions on Computers* 70.1 (2020), pp. 17–29.

[FL13] S. Franey and M. Lipasti. "Accelerating atomic operations on GPGPUs." In: *NoCS.* 2013.

[Fla+18] Eric Flamand, Davide Rossi, Francesco Conti, Igor Loi, Antonio Pullini, Florent Rotenberg, and Luca Benini. "GAP-8: A RISC-V SoC for AI at the Edge of the IoT." In: *2018 IEEE 29th International Conference on Application-specific Systems, Architectures and Processors (ASAP).* 2018, pp. 1–4. DOI: 10.1109/ASAP.2018.8445101.

[For+20] Björn Forsberg, Maxim Mattheeuws, Andreas Kurth, Andrea Marongiu, and Luca Benini. "A Synergistic Approach to Predictable Compilation and Scheduling on Commodity Multi-Cores." In: *The 21st ACM SIGPLAN/SIGBED Conference on Languages, Compilers, and Tools for Embedded Systems.* LCTES '20. London, United Kingdom: Association for Computing Machinery, 2020, pp. 108–118. ISBN: 9781450370943. DOI: 10.1145/3372799.3394369. URL: https://doi.org/10.1145/3372799.3394369.

[Fru16] Andrei Frumusanu. "Cortex A72 Performance and Power." In: *AnandTech* (2016).

[Fru18] Andrei Frumusanu. "The Qualcomm Snapdragon 855 Pre-Dive: Going Into Detail on 2019's Flagship Android SoC." In: *AnandTech* (Dec. 2018). URL: https://www.anandtech.com/show/13680/snapdragon-855-going-into-detail.

[Fru20a] Andrei Frumusanu. "Apple Announces 5nm A14 SoC – Meagre Upgrades, Or Just Less Power Hungry?" In: *AnandTech* (Sept. 2020). URL: https://www.anandtech.com/show/16088.

[Fru20b] Andrei Frumusanu. "Qualcomm Details the Snapdragon 888: 3rd Gen 5G & Cortex-X1 on 5nm." In: *AnandTech* (Dec. 2020). URL: https://www.anandtech.com/show/16271.

[Fru20c] Andrei Frumusanu. "The 2020 Mac Mini Unleashed: Putting Apple Silicon M1 To The Test." In: *AnandTech* (Nov. 2020). URL: https://www.anandtech.com/show/16252.

[Fun20] Fungible Inc. "The Fungible DPU: A New Category of Microprocessor for the Data-Centric Era." In: *2020 IEEE Hot Chips 32 Symposium (HCS)*. Los Alamitos, CA, USA: IEEE Computer Society, Aug. 2020, pp. 1–25. DOI: 10.1109/HCS49909.2020.9220423. URL: https://doi.ieeecomputersociety.org/10.1109/HCS49909.2020.9220423.

[FW86] Philip J. Fleming and John J. Wallace. "How Not to Lie with Statistics: The Correct Way to Summarize Benchmark Results." In: *Commun. ACM* 29.3 (Mar. 1986), pp. 218–221. ISSN: 0001-0782. DOI: 10.1145/5666.5673. URL: https://doi.org/10.1145/5666.5673.

[Gau+17] Michael Gautschi, Pasquale Davide Schiavone, Andreas Traber, Igor Loi, Antonio Pullini, Davide Rossi, Eric Flamand, Frank K. Gürkaynak, and Luca Benini. "Near-Threshold RISC-V Core With DSP Extensions for Scalable IoT Endpoint Devices." In: *IEEE Transactions on Very Large Scale Integration (VLSI)*

Systems 25.10 (2017), pp. 2700–2713. DOI: 10.1109/TVLSI.2017.2654506.

[GB06] I. Ganusov and M. Burtscher. "Efficient emulation of hardware prefetchers via event-driven helper threading." In: *2006 International Conference on Parallel Architectures and Compilation Techniques (PACT)*. Sept. 2006, pp. 144–153.

[GGL12] Tobias Grosser, Armin Groesslinger, and Christian Lengauer. "Polly—performing polyhedral optimizations on a low-level intermediate representation." In: *Parallel Processing Letters* 22.04 (2012), p. 1250010.

[GH16] Tobias Grosser and Torsten Hoefler. "Polly-ACC Transparent Compilation to Heterogeneous Hardware." In: *Proceedings of the 2016 International Conference on Supercomputing*. ICS '16. Istanbul, Turkey: ACM, 2016, 1:1–1:13. ISBN: 978-1-4503-4361-9. DOI: 10.1145/2925426.2926286. URL: http://doi.acm.org/10.1145/2925426.2926286.

[Gha+90] Kourosh Gharachorloo, Daniel Lenoski, James Laudon, Phillip Gibbons, Anoop Gupta, and John Hennessy. "Memory consistency and event ordering in scalable shared-memory multiprocessors." In: *ISCA*. 1990.

[Gir+21] Davide Giri, Kuan-Lin Chiu, Guy Eichler, Paolo Mantovani, and Luca P. Carloni. "Accelerator Integration for Open-Source SoC Design." In: *IEEE Micro* 41.4 (2021), pp. 8–14. DOI: 10.1109/MM.2021.3073893.

[GMC18a] Davide Giri, Paolo Mantovani, and Luca P. Carloni. "NoC-Based Support of Heterogeneous Cache-Coherence Models for Accelerators." In: *IEEE/ACM NOCS*. 2018.

[GMC18b] D. Giri, P. Mantovani, and L. P. Carloni. "Accelerators and Coherence: An SoC Perspective." In: *IEEE Micro* (2018).

[Gom+13] J. Gomez-Luna, J. M. Gonzalez-Linares, J. I. Benavides Benitez, and N. Guil Mata. "Performance Modeling of Atomic Additions on GPU Scratchpad Memory." In: *IEEE TPDS* (2013).

[Goo13] J. Goodacre. "The Evolution of the ARM Architecture Towards Big Data and the Data-centre." In: *VHPC '13*. VHPC '13. Denver, Colorado: ACM, 2013, 4:1–4:1. ISBN: 978-1-4503-2509-7. DOI: 10.1145/2535800.2535921. URL: http://doi.acm.org/10.1145/2535800.2535921.

[Got+83] A. Gottlieb, R. Grishman, C. Kruskal, K. McAuliffe, L. Rudolph, and M. Snir. "The NYU Ultracomputer—Designing an MIMD Shared Memory Parallel Computer." In: *IEEE TC* (1983).

[Gra+12] Scott Grauer-Gray, Lifan Xu, Robert Searles, Sudhee Ayalasomayajula, and John Cavazos. "Auto-tuning a high-level language targeted to GPU codes." In: *2012 Innovative Parallel Computing (InPar)*. IEEE. 2012, pp. 1–10.

[Gra16] J. Gray. "GRVI Phalanx: A Massively Parallel RISC-V FPGA Accelerator Accelerator." In: *2016 IEEE 24th Annual International Symposium on Field-Programmable Custom Computing Machines (FCCM)*. May 2016, pp. 17–20. DOI: 10.1109/FCCM.2016.12.

[Gra19] J. Gray. "2GRVI Phalanx: Towards Kilocore RISC-V FPGA Accelerators with HBM2 DRAM." In: *IEEE/ACM HotChips 30*. 2019.

[Gui+19] Chuang-Yi Gui, Long Zheng, Bingsheng He, Cheng Liu, Xin-Yu Chen, Xiao-Fei Liao, and Hai Jin. "A survey on graph processing accelerators: Challenges and opportunities." In: *Journal of Computer Science and Technology* 34.2 (2019), pp. 339–371.

[Guo+14] Y. Guo, M. Biczak, A. L. Varbanescu, A. Iosup, C. Martella, and T. L. Willke. "How Well do Graph-Processing Platforms Perform? An Empirical Performance Evaluation and Analysis." In: *IEEE IPDPS '14*. May 2014, pp. 395–404.

[Hao+17] Yuchen Hao, Zhenman Fang, Glenn Reinman, and Jason Cong. "Supporting Address Translation for Accelerator-Centric Architectures." In: *2017 IEEE International Symposium on High Performance Computer Architecture (HPCA)*. 2017, pp. 37–48. DOI: 10.1109/HPCA.2017.19.

[Her14] Jörg Herter. "Timing-Predictable Memory Allocation in Hard Real-Time Systems." PhD thesis. Universität des Saarlandes, 2014.

[Hew21] Hewlett Packard Enterprise. *HPE Cray Supercomputers*. Data Sheet. Apr. 2021. URL: https://www.hpe.com/psnow/doc/PSN1012927320USEN.pdf.

[HM08] M. D. Hill and M. R. Marty. "Amdahl's Law in the Multicore Era." In: *Computer* (2008).

[Hor14] Mark Horowitz. "Computing's energy problem (and what we can do about it)." In: *2014 IEEE International Solid-State Circuits Conference (ISSCC)*. 2014, pp. 10–14. DOI: 10.1109/ISSCC.2014.6757323.

[HP17] J.L. Hennessy and D.A. Patterson. In: *Computer Architecture: A Quantitative Approach*. The Morgan Kaufmann Series in Computer Architecture and Design. Elsevier Science, 2017. Chap. 7.2 Guidelines for Domain-Specific Architectures. ISBN: 9780128119051.

[HP19] John L. Hennessy and David A. Patterson. "A New Golden Age for Computer Architecture." In: *Commun. ACM* 62.2 (Jan. 2019), pp. 48–60. ISSN: 0001-0782. DOI: 10.1145/3282307. URL: https://doi.org/10.1145/3282307.

[HS11] Maurice Herlihy and Nir Shavit. *The Art of Multiprocessor Programming*. Morgan Kaufmann, 2011.

[HSA12] HSA Foundation. *HSA Foundation*. www.hsafoundation.com. June 2012.

[HTE17] Simon David Hammond, Christian Robert Trott, and Harold C Edwards. *On the Importance of Faster Atomics*. 2017.

[Hua+19] Qijing Huang, Christopher Yarp, Sagar Karandikar, Nathan Pemberton, Benjamin Brock, Liang Ma, Guohao Dai, Robert Quitt, Krste Asanovic, and John Wawrzynek. "Centrifuge: Evaluating full-system HLS-generated heterogenous-accelerator SoCs using FPGA-Acceleration." In: *2019 IEEE/ACM International Conference on Computer-Aided Design (ICCAD)*. 2019, pp. 1–8. DOI: 10.1109/ICCAD45719.2019.8942048.

[Hua+21] Bo-Yuan Huang, Steven Lyubomirsky, Thierry Tambe, Yi Li, Mike He, Gus Smith, Gu-Yeon Wei, Aarti Gupta, Sharad Malik, and Zachary Tatlock. "From DSLs to Accelerator-Rich Platform Implementations: Addressing the Mapping Gap." In: (2021).

[Hwu16] W.-M. W. Hwu, ed. *Heterogeneous System Architecture*. Morgan Kaufmann Publishers, 2016. ISBN: 978-0-12-800386-2.

[IBM07] *CoreConnect Processor Local Bus Specification*. IBM Inc. 2007.

[Int15] Intel Corp. *The compute architecture of Intel Processor Graphics Gen9*. White Paper. 2015. URL: https://software.intel.com/sites/default/files/managed/c5/9a/The-Compute-Architecture-of-Intel-Processor-Graphics-Gen9-v1d0.pdf.

[Int18] *Migrating Offloading Software to Intel Xeon Phi Processor*. White Paper. Intel Corp., Feb. 2018. URL: https://www.intel.cn/content/dam/www/public/us/en/documents/white-papers/migrating-offloading-software-paper.pdf.

[Int19a] Intel Corporation. *Intel 64 and IA-32 architectures optimization reference manual*. 2019.

[Int19b] Intel Corporation. *Intel 64 and IA-32 architectures software developer's manual.* 2019.

[Int20a] Intel Corp. *Intel Movidius Myriad X Vision Processing Units.* Data Book. 2020. URL: https://cdrdv2.intel.com/v1/dl/getContent/621494.

[Int20b] *oneAPI Specification, Release 1.0-rev-3.* Intel Corp. Nov. 2020.

[Jan+11] Byunghyun Jang, Dana Schaa, Perhaad Mistry, and David Kaeli. "Exploiting Memory Access Patterns to Improve Memory Performance in Data-Parallel Architectures." In: *IEEE Transactions on Parallel and Distributed Systems* 22.1 (2011), pp. 105–118. DOI: 10.1109/TPDS.2010.107.

[Jer+17] N.E. Jerger, T. Krishna, L.S. Peh, and M. Martonosi. *On-Chip Networks: Second Edition.* Morgan & Claypool, 2017. ISBN: 9781627059961.

[JHL20] Y. Jiao, L. Han, and X. Long. "Hanguang 800 NPU: The Ultimate AI Inference Solution for Data Centers." In: *2020 IEEE Hot Chips 32 Symposium (HCS).* Los Alamitos, CA, USA: IEEE Computer Society, Aug. 2020, pp. 1–29. DOI: 10.1109/HCS49909.2020.9220619. URL: https://doi.ieeecomputersociety.org/10.1109/HCS49909.2020.9220619.

[Jia+18] Zhe Jia, Marco Maggioni, Benjamin Staiger, and Daniele P Scarpazza. *Dissecting the NVIDIA Volta GPU architecture via microbenchmarking.* 2018.

[Joh+11] D. Johnson et al. "Rigel: A 1,024-Core Single-Chip Accelerator Architecture." In: *IEEE Micro* 31.4 (July 2011), pp. 30–41. ISSN: 0272-1732. DOI: 10.1109/MM.2011.40.

[Jou+17] N. P. Jouppi et al. "In-datacenter performance analysis of a tensor processing unit." In: *2017 ACM/IEEE 44th Annual International Symposium on Computer Architecture (ISCA).* June 2017, pp. 1–12. DOI: 10.1145/3079856.3080246.

[Jou+18] N. Jouppi, C. Young, N. Patil, and D. Patterson. "Motivation for and Evaluation of the First Tensor Processing Unit." In: *IEEE Micro* 38.3 (May 2018), pp. 10–19. DOI: 10.1109/MM.2018.032271057.

[KAH18] Hadi Mardani Kamali, Kimia Zamiri Azar, and Shaahin Hessabi. "DuCNoC: A High-Throughput FPGA-Based NoC Simulator Using Dual-Clock Lightweight Router Micro-Architecture." In: *IEEE Transactions on Computers* 67.2 (2018), pp. 208–221. DOI: 10.1109/TC.2017.2735399.

[Kal14] Kalray S.A. *MPPA MANYCORE*. Feb. 2014.

[Kal20] Kalray S.A. *MPPA3 Coolidge Processor*. White Paper. Jan. 2020. URL: https://www.kalrayinc.com/documentation/.

[Kan16] D. Kanter. "RISC-V Offers Simple, Modular ISA: New CPU Instruction Set Is Open and Extensible." In: *Microprocessor Report* (Mar. 2016). URL: https://riscv.org/2016/04/risc-v-offers-simple-modular-isa.

[Kar+18] Sagar Karandikar et al. "FireSim: FPGA-Accelerated Cycle-Exact Scale-Out System Simulation in the Public Cloud." In: *2018 ACM/IEEE 45th Annual International Symposium on Computer Architecture (ISCA)*. 2018, pp. 29–42. DOI: 10.1109/ISCA.2018.00014.

[Kar+20] Sagar Karandikar, Albert Ou, Alon Amid, Howard Mao, Randy Katz, Borivoje Nikolić, and Krste Asanović. "FirePerf: FPGA-Accelerated Full-System Hardware/-Software Performance Profiling and Co-Design." In: *Proceedings of the Twenty-Fifth International Conference on Architectural Support for Programming Languages and Operating Systems*. ASPLOS '20. Lausanne, Switzerland: Association for Computing Machinery, 2020, pp. 715–731. ISBN: 9781450371025. DOI: 10.1145/3373376.3378455. URL: https://doi.org/10.1145/3373376.3378455.

[KC18] Santanu Kundu and Santanu Chattopadhyay. *Network-on-Chip: The Next Generation of System-on-Chip Integration.* CRC Press, 2018. ISBN: 9781466565272.

[KDS00] M. Karlsson, F. Dahlgren, and P. Stenstrom. "A prefetching technique for irregular accesses to linked data structures." In: *Proceedings Sixth International Symposium on High-Performance Computer Architecture. HPCA-6 (Cat. No.PR00550).* 2000, pp. 206–217. DOI: 10.1109/HPCA.2000.824351.

[Ker10] M. Kerrisk. *The Linux Programming Interface: A Linux and UNIX System Programming Handbook.* No Starch Press, 2010. ISBN: 9781593272913.

[KFB22] Andreas Kurth, Björn Forsberg, and Luca Benini. "HEROv2: Full-Stack Open-Source Research Platform for Heterogeneous Computing." In: *IEEE Transactions on Parallel and Distributed Systems* (2022). DOI: 10.1109/TPDS.2022.3189390.

[KHG20] A. Kamaleldin, S. Hesham, and D. Göhringer. "Towards a Modular RISC-V Based Many-Core Architecture for FPGA Accelerators." In: *IEEE Access* 8 (2020), pp. 148812–148826. DOI: 10.1109/ACCESS.2020.3015706.

[Khr19] Khronos Group Inc. *OpenVX API Specification 1.3.* Aug. 2019. URL: https://www.khronos.org/registry/OpenVX/specs/1.3/OpenVX_Specification_1_3.pdf.

[Khr21a] *SYCL 2020 Specification (revision 3).* Khronos SYCL Working Group. Mar. 2021.

[Khr21b] *The OpenCL C Specification v3.0.7.* Khronos OpenCL Working Group. Apr. 2021.

[Kir20] Pavel Kirienko. *Constant-complexity deterministic memory allocator (heap) for hard real-time high-integrity embedded systems.* 2020. URL: https://github.com/pavel-kirienko/o1heap.

[Kli20] A. A. Klimentov. "Exascale Data Processing in Heterogeneous Distributed Computing Infrastructure for Applications in High Energy Physics." In: *Physics of Particles and Nuclei* (Nov. 2020). DOI: 10.1134/S1063779620060052.

[kok18] kokke. *tiny-AES-c: Small portable AES 128/192/256 in C*. 2018. URL: https://github.com/kokke/tiny-AES-c/tree/f56dbc05ab0d795d74f43436aac9da56a7cc8e11.

[Kor+14] G. Kornaros, K. Harteros, I. Christoforakis, and M. Astrinaki. "I/O Virtualization Utilizing an Efficient Hardware System-Level Memory Management Unit." In: *ISSoC '14*. 2014, pp. 1–4.

[KS02] G. B. Kandiraju and A. Sivasubramaniam. "Going the Distance for TLB Prefetching: An Application-driven Study." In: *Proceedings of the 29th Annual International Symposium on Computer Architecture*. ISCA '02. Anchorage, Alaska: IEEE Computer Society, 2002, pp. 195–206. ISBN: 0-7695-1605-X. URL: http://dl.acm.org/citation.cfm?id=545215.545237.

[Kur+16] Andreas Kurth, Andreas Tretter, Pascal A. Hager, Sergio Sanabria, Orcun Goksel, Lothar Thiele, and Luca Benin. "Mobile ultrasound imaging on heterogeneous multi-core platforms." In: *2016 14th ACM/IEEE Symposium on Embedded Systems For Real-time Multimedia (ESTIMedia)*. 2016, pp. 1–10.

[Kur+17] Andreas Kurth, Pirmin Vogel, Alessandro Capotondi, Andrea Marongiu, and Luca Benini. "HERO: Heterogeneous Embedded Research Platform for Exploring RISC-V Manycore Accelerators on FPGA." In: *Proceedings of the First International Workshop on Computer Architecture Research with RISC-V*. CARRV '17. Oct. 2017. DOI: 10.3929/ethz-b-000219249.

[Kur+18a] A. Kurth, P. Vogel, A. Marongiu, and L. Benini. "Scalable and Efficient Virtual Memory Sharing in Heterogeneous SoCs with TLB Prefetching and MMU-Aware DMA Engine." In: *2018 IEEE 36th International Conference on Computer Design (ICCD)*. Oct. 2018, pp. 292–300. DOI: 10.1109/ICCD.2018.00052.

[Kur+18b] Andreas Kurth, Alessandro Capotondi, Pirmin Vogel, Luca Benini, and Andrea Marongiu. "HERO: An Open-Source Research Platform for HW/SW Exploration of Heterogeneous Manycore Systems." In: *Proceedings of the 2nd Workshop on AutotuniNg and aDaptivity AppRoaches for Energy Efficient HPC Systems*. ANDARE '18. Limassol, Cyprus: ACM, Nov. 2018, 5:1–5:6. ISBN: 978-1-4503-6591-8. DOI: 10.1145/3295816.3295821. URL: http://doi.acm.org/10.1145/3295816.3295821.

[Kur+18c] Andreas Kurth, Alessandro Capotondi, Pirmin Vogel, Luca Benini, and Andrea Marongiu. "HERO: An Open-Source Research Platform for HW/SW Exploration of Heterogeneous Manycore Systems." In: *Proceedings of the 2nd Workshop on AutotuniNg and ADaptivity AppRoaches for Energy Efficient HPC Systems*. ANDARE '18. Limassol, Cyprus: Association for Computing Machinery, 2018. ISBN: 9781450365918. DOI: 10.1145/3295816.3295821. URL: https://doi.org/10.1145/3295816.3295821.

[Kur+18d] Andreas Kurth, Pirmin Vogel, Andrea Marongiu, and Luca Benini. "Scalable and Efficient Virtual Memory Sharing in Heterogeneous SoCs with TLB Prefetching and MMU-Aware DMA Engine." In: *2018 IEEE 36th International Conference on Computer Design (ICCD)*. 2018, pp. 292–300. DOI: 10.1109/ICCD.2018.00052.

[Kur+20a] Andreas Kurth, Samuel Riedel, Florian Zaruba, Torsten Hoefler, and Luca Benini. "ATUNs: Modular and Scalable Support for Atomic Operations in a Shared Memory Multiprocessor." In: *Proceedings of the 57th*

ACM/IEEE Design Automation Conference. DAC 2020. San Francisco, CA, USA, June 2020, pp. 1–6. DOI: 10.1109/DAC18072.2020.9218661.

[Kur+20b] Andreas Kurth, Koen Wolters, Björn Forsberg, Alessandro Capotondi, Andrea Marongiu, Tobias Grosser, and Luca Benini. "Mixed-Data-Model Heterogeneous Compilation and OpenMP Offloading." In: *Proceedings of the 29th International Conference on Compiler Construction.* CC 2020. San Diego, CA, USA: Association for Computing Machinery, Feb. 2020, pp. 119–131. ISBN: 9781450371209. DOI: 10.1145/3377555.3377891. URL: https://doi.org/10.1145/3377555.3377891.

[Kur+22] Andreas Kurth, Wolfgang Rönninger, Thomas Benz, Matheus Cavalcante, Fabian Schuiki, Florian Zaruba, and Luca Benini. "An Open-Source Platform for High-Performance Non-Coherent On-Chip Communication." In: *IEEE Transactions on Computers* 71.8 (2022), pp. 1794–1809. DOI: 10.1109/TC.2021.3107726.

[LAC14] M. Lavasani, H. Angepat, and D. Chiou. "An FPGA-based In-Line Accelerator for Memcached." In: *IEEE CAL* 13.2 (July 2014), pp. 57–60.

[LAG11] H. J. Lu, H. Peter Anvin, and Milind Girkar. "X32: A native 32-bit ABI for x86-64." In: *Linux Plumbers Conference.* 2011. URL: http://www.linuxplumbersconf.net/2011/ocw/system/presentations/531/original/x32-LPC-2011-0906.pptx.

[Lat+21a] Chris Lattner, Mehdi Amini, Uday Bondhugula, Albert Cohen, Andy Davis, Jacques Pienaar, River Riddle, Tatiana Shpeisman, Nicolas Vasilache, and Oleksandr Zinenko. "MLIR: Scaling Compiler Infrastructure for Domain Specific Computation." In: *2021 IEEE/ACM International Symposium on Code Generation and Optimization (CGO).* Feb. 2021, pp. 2–14. DOI: 10.1109/CGO51591.2021.9370308.

[Lat+21b] Chris Lattner, John Demme, Mike Urbach, Andrew Lenharth, Schuyler Eldridge, Andrew Young, Fabian Schuiki, and Stephen Neuendorffer. *CIRCT: Circuit IR Compilers and Tools.* June 2021. URL: https://circt.llvm.org/.

[LBM13] D. Lustig, A. Bhattacharjee, and M. Martonosi. "TLB Improvements for Chip Multiprocessors: Inter-Core Cooperative Prefetchers and Shared Last-Level TLBs." In: *ACM Transactions on Architecture and Code Optimization (TACO)* 10.1 (Apr. 2013), 2:1–2:38. ISSN: 1544-3566. DOI: 10.1145/2445572.2445574. URL: http://doi.acm.org/10.1145/2445572.2445574.

[Lee+09] J. Lee, C. Jung, D. Lim, and Y. Solihin. "Prefetching with Helper Threads for Loosely Coupled Multiprocessor Systems." In: *IEEE Transactions on Parallel and Distributed Systems* 20.9 (Sept. 2009), pp. 1309–1324. ISSN: 1045-9219. DOI: 10.1109/TPDS.2008.224.

[Lee+16] Y. Lee et al. "An Agile Approach to Building RISC-V Microprocessors." In: *IEEE Micro* 36.2 (2016), pp. 8–20. DOI: 10.1109/MM.2016.11.

[Li+09] Sheng Li, Jung Ho Ahn, Richard D. Strong, Jay B. Brockman, Dean M. Tullsen, and Norman P. Jouppi. "McPAT: An Integrated Power, Area, and Timing Modeling Framework for Multicore and Manycore Architectures." In: *Proceedings of the 42nd Annual IEEE/ACM International Symposium on Microarchitecture.* MICRO 42. New York, New York: Association for Computing Machinery, 2009, pp. 469–480. ISBN: 9781605587981. DOI: 10.1145/1669112.1669172. URL: https://doi.org/10.1145/1669112.1669172.

[Lia+19] H. Liao, J. Tu, J. Xia, and X. Zhou. "DaVinci: A Scalable Architecture for Neural Network Computing." In: *2019 IEEE Hot Chips 31 Symposium (HCS).* Los Alamitos, CA, USA: IEEE Computer Society, Aug. 2019, pp. 1–44. DOI: 10.1109/HOTCHIPS.2019.

8875654. URL: https://doi.ieeecomputersociety.org/10.1109/HOTCHIPS.2019.8875654.

[Lit01] Tim Littlefair. "An investigation into the use of software code metrics in the industrial software development environment." PhD thesis. Edith Cowan University, 2001.

[Liu+18] Yang Liu, Haiwei Dong, Longyu Zhang, and Abdulmotaleb El Saddik. "Technical Evaluation of HoloLens for Multimedia: A First Look." In: *IEEE MultiMedia* 25.4 (2018), pp. 8–18. DOI: 10.1109/MMUL.2018.2873473.

[LL03] Marisa López-Vallejo and Juan Carlos López. "On the Hardware-Software Partitioning Problem: System Modeling and Partitioning Techniques." In: *ACM Trans. Des. Autom. Electron. Syst.* 8.3 (July 2003), pp. 269–297. ISSN: 1084-4309. DOI: 10.1145/785411.785412. URL: https://doi.org/10.1145/785411.785412.

[LLV19] LLVM. *LLVM 8.0.0 Release Notes.* Mar. 2019. URL: https://releases.llvm.org/8.0.0/docs/ReleaseNotes.html.

[LM99] C.-K. Luk and T. C. Mowry. "Automatic compiler-inserted prefetching for pointer-based applications." In: *IEEE Transactions on Computers* 48.2 (Feb. 1999), pp. 134–141. ISSN: 0018-9340. DOI: 10.1109/12.752654.

[Loh10] Eugene Loh. "The Ideal HPC Programming Language." In: *Commun. ACM* 53.7 (July 2010), pp. 42–47. ISSN: 0001-0782. DOI: 10.1145/1785414.1785433. URL: https://doi.org/10.1145/1785414.1785433.

[Loi+15] I. Loi, D. Rossi, G. Haugou, M. Gautschi, and L. Benini. "Exploring Multi-banked shared-L1 Program Cache on Ultra-low Power, Tightly Coupled Processor Clusters." In: *Proceedings of the 12th ACM International Conference on Computing Frontiers.* CF '15.

Ischia, Italy: ACM, 2015, 64:1–64:8. ISBN: 978-1-4503-3358-0. DOI: 10.1145/2742854.2747288. URL: http://doi.acm.org/10.1145/2742854.2747288.

[Loi+18] I. Loi, A. Capotondi, D. Rossi, A. Marongiu, and L. Benini. "The Quest for Energy-Efficient I$ Design in Ultra-Low-Power Clustered Many-Cores." In: *IEEE Transactions on Multi-Scale Computing Systems* 4.2 (Apr. 2018), pp. 99–112. ISSN: 2332-7766. DOI: 10.1109/TMSCS.2017.2769046.

[LS94] A. R. Lebeck and G. S. Sohi. "Request combining in multiprocessors with arbitrary interconnection networks." In: *IEEE TPDS* (1994).

[Luk01] C.-K. Luk. "Tolerating Memory Latency Through Software-controlled Pre-execution in Simultaneous Multithreading Processors." In: *Proceedings of the 28th Annual International Symposium on Computer Architecture.* ISCA '01. Göteborg, Sweden: ACM, 2001, pp. 40–51. ISBN: 0-7695-1162-7. DOI: 10.1145/379240.379250. URL: http://doi.acm.org/10.1145/379240.379250.

[Lut18] Andy Lutomirski. *Can we drop upstream Linux x32 support?* 2018. URL: https://lkml.org/lkml/2018/12/10/1145.

[LW17] Shin-Ying Lee and Carole-Jean Wu. "Performance characterization, prediction, and optimization for heterogeneous systems with multi-level memory interference." In: *2017 IEEE International Symposium on Workload Characterization (IISWC).* 2017, pp. 43–53. DOI: 10.1109/IISWC.2017.8167755.

[Mac+21] Stefan Mach, Fabian Schuiki, Florian Zaruba, and Luca Benini. "FPnew: An Open-Source Multiformat Floating-Point Unit Architecture for Energy-Proportional Transprecision Computing." In: *IEEE Transactions on Very Large Scale Integration (VLSI) Systems* 29.4 (2021), pp. 774–787. DOI: 10.1109/TVLSI.2020.3044752.

[Man+20] P. Mantovani, D. Giri, G. Di Guglielmo, L. Piccolboni, J. Zuckerman, E. G. Cota, M. Petracca, C. Pilato, and L. P. Carloni. "Agile SoC Development with Open ESP." In: *2020 IEEE/ACM International Conference On Computer Aided Design (ICCAD)*. 2020, pp. 1–9.

[Mar+15] A. Marongiu, A. Capotondi, G. Tagliavini, and L. Benini. "Simplifying Many-Core-Based Heterogeneous SoC Programming With Offload Directives." In: *IEEE Transactions on Industrial Informatics* 11.4 (Aug. 2015), pp. 957–967. ISSN: 1551-3203. DOI: 10.1109/TII.2015.2449994.

[Mar+16] Matt Martineau, Simon McIntosh-Smith, Carlo Bertolli, Arpith C Jacob, Samuel F Antao, Alexandre Eichenberger, Gheorghe-Teodor Bercea, Tong Chen, Tian Jin, Kevin O'Brien, et al. "Performance analysis and optimization of Clang's OpenMP 4.5 GPU support." In: *2016 7th International Workshop on Performance Modeling, Benchmarking and Simulation of High Performance Computer Systems (PMBS)*. IEEE. 2016, pp. 54–64.

[Mar+18] Romolo Marotta, Mauro Ianni, Alessandro Pellegrini, Andrea Scarselli, and Francesco Quaglia. *A Non-blocking Buddy System for Scalable Memory Allocation on Multi-core Machines*. 2018.

[Mat+21] Maxim Mattheeuws, Björn Forsberg, Andreas Kurth, and Luca Benini. "Analyzing memory interference of FPGA accelerators on multicore hosts in heterogeneous reconfigurable SoCs." In: *Proceedings of the 2021 IEEE/ACM Design, Automation and Test in Europe Conference*. DATE '21. Mar. 2021.

[MCB16] Andrea Marongiu, Alessandro Capotondi, and Luca Benini. "Controlling NUMA effects in embedded manycore applications with lightweight nested parallelism support." In: *Parallel Computing* 59 (2016). Theory and Practice of Irregular Applications, pp. 24–42. ISSN: 0167-8191. DOI: https://doi.org/

10.1016/j.parco.2016.02.002. URL: https://www.sciencedirect.com/science/article/pii/S0167819116000442.

[Mel+12] D. Melpignano et al. "Platform 2012, a Many-core Computing Accelerator for Embedded SoCs: Performance Evaluation of Visual Analytics Applications." In: *Proceedings of the 49th Annual Design Automation Conference*. DAC '12. San Francisco, California: ACM, 2012, pp. 1137–1142. ISBN: 978-1-4503-1199-1. DOI: 10.1145/2228360.2228568. URL: http://doi.acm.org/10.1145/2228360.2228568.

[Men19] Mentor, a Siemens Business. *Questa Advanced Simulator*. 2019. URL: https://www.mentor.com/products/fv/questa/.

[Mic18] *C++ Accelerated Massive Parallelism (C++ AMP)*. Microsoft Corp. Nov. 2018.

[Mik+14] Dmitry Mikushin, Nikolay Likhogrud, Eddy Z Zhang, and Christopher Bergström. "KernelGen–The Design and Implementation of a Next Generation Compiler Platform for Accelerating Numerical Models on GPUs." In: *2014 IEEE International Parallel & Distributed Processing Symposium Workshops*. IEEE. 2014, pp. 1011–1020.

[Mit+14] Gaurav Mitra, Eric Stotzer, Ajay Jayaraj, and Alistair P. Rendell. "Implementation and Optimization of the OpenMP Accelerator Model for the TI Keystone II Architecture." In: *Using and Improving OpenMP for Devices, Tasks, and More*. Ed. by Luiz DeRose, Bronis R. de Supinski, Stephen L. Olivier, Barbara M. Chapman, and Matthias S. Müller. Cham: Springer International Publishing, 2014, pp. 202–214. ISBN: 978-3-319-11454-5.

[MMG16] M. Martineau, S. McIntosh-Smith, and W. Gaudin. "Evaluating OpenMP 4.0's Effectiveness as a Heterogeneous Parallel Programming Model." In: *2016*

IEEE International Parallel and Distributed Processing Symposium Workshops (IPDPSW). May 2016, pp. 338–347. DOI: 10.1109/IPDPSW.2016.70.

[Moo19] Samuel K. Moore. "Intel's View of the Chiplet Revolution." In: *IEEE Spectrum* (Apr. 2019).

[Mor+14] Amir Morad, Tomer Y. Morad, Yavits Leonid, Ran Ginosar, and Uri Weiser. "Generalized MultiAmdahl: Optimization of Heterogeneous Multi-Accelerator SoC." In: *IEEE Computer Architecture Letters* 13.1 (2014), pp. 37–40. DOI: 10.1109/L-CA.2012.34.

[Mul17] David Mulnix. *Intel Xeon Processor Scalable Family Technical Overview.* Intel Corp., 2017.

[MV15] Sparsh Mittal and Jeffrey S. Vetter. "A Survey of CPU-GPU Heterogeneous Computing Techniques." In: *ACM Comput. Surv.* 47.4 (July 2015). ISSN: 0360-0300. DOI: 10.1145/2788396. URL: https://doi.org/10.1145/2788396.

[Nab+20] Seyed Morteza Nabavinejad, Mohammad Baharloo, Kun-Chih Chen, Maurizio Palesi, Tim Kogel, and Masoumeh Ebrahimi. "An Overview of Efficient Interconnection Networks for Deep Neural Network Accelerators." In: *IEEE JESTCS* 10.3 (2020), pp. 268–282. DOI: 10.1109/JETCAS.2020.3022920.

[Nor+20] T. Norrie, N. Patil, D. Yoon, G. Kurian, S. Li, J. Laudon, C. Young, N. P. Jouppi, and D. Patterson. "Google's Training Chips Revealed: TPUv2 and TPUv3." In: *2020 IEEE Hot Chips 32 Symposium (HCS).* Los Alamitos, CA, USA: IEEE Computer Society, Aug. 2020, pp. 1–70. DOI: 10.1109/HCS49909.2020.9220735. URL: https://doi.ieeecomputersociety.org/10.1109/HCS49909.2020.9220735.

[Nvi14] Nvidia Corp. *Summit and Sierra Supercomputers: An Inside Look at the U.S. Department of Energy's New Pre-Exascale Systems.* Nov. 2014. URL: http://

www.teratec.eu/actu/calcul/Nvidia_Coral_White_Paper_Final_3_1.pdf.

[Nvi15] Nvidia Corp. *Linux for Tegra R23.1*. Nov. 2015. URL: https://developer.nvidia.com/embedded/linux-tegra-r231.

[Nvi16] Nvidia Corp. *Linux for Tegra R24.1*. May 2016. URL: https://developer.nvidia.com/embedded/linux-tegra-r241.

[Nvi17] *Nvidia Tesla V100 GPU Architecture: The World's Most Advanced Data Center GPU*. White Paper. Nvidia Corp., Aug. 2017. URL: https://images.nvidia.com/content/volta-architecture/pdf/volta-architecture-whitepaper.pdf.

[Nvi19a] Nvidia Corp. *Nvidia TITAN RTX*. Product Brief. Apr. 2019. URL: https://www.nvidia.com/content/dam/en-zz/Solutions/titan/documents/titan-rtx-for-creators-us-nvidia-1011126-r6-web.pdf.

[Nvi19b] Nvidia Corp. *NVVM IR Specification 1.5*. Oct. 2019. URL: https://docs.nvidia.com/cuda/nvvm-ir-spec/index.html.

[Nvi20a] *NVIDIA Jetson AGX Xavier Module Data Sheet*. Nvidia Corp. 2020. URL: https://developer.nvidia.com/embedded/dlc/jetson-xavier-data-sheet.

[Nvi20b] *NVIDIA Jetson NX Module Data Sheet*. Nvidia Corp. 2020. URL: hhttps://developer.nvidia.com/jetson-xavier-nx-data-sheet.

[Ogu+10] T. Oguntebi et al. "FARM: A Prototyping Environment for Tightly-Coupled, Heterogeneous Architectures." In: *2010 18th IEEE Annual International Symposium on Field-Programmable Custom Computing Machines*. FCCM '10. May 2010, pp. 221–228. DOI: 10.1109/FCCM.2010.41.

[Oh17] Nate Oh. "Intel Announces Movidius Myriad X VPU, Featuring 'Neural Compute Engine'." In: *AnandTech* (Aug. 2017). URL: https://www.anandtech.com/show/11771/intel-announces-movidius-myriad-x-vpu.

[Olo16] A. Olofsson. "Epiphany-V: A 1024 processor 64-bit RISC System-On-Chip." In: *CoRR* abs/1610.01832 (Oct. 2016). URL: http://arxiv.org/abs/1610.01832.

[Omp3] *OpenMP Application Programming Interface, Version 3.1.* OpenMP Architecture Review Board. July 2011. URL: https://www.openmp.org/wp-content/uploads/OpenMP3.1.pdf.

[Omp4.0] *OpenMP Application Programming Interface, Version 4.0.* OpenMP Architecture Review Board. July 2013. URL: https://www.openmp.org/wp-content/uploads/OpenMP4.0.0.pdf.

[Omp4.5] *OpenMP Application Programming Interface, Version 4.5.* OpenMP Architecture Review Board. 2015.

[Omp5.1] *OpenMP Application Programming Interface, Version 5.1.* OpenMP Architecture Review Board. Nov. 2020. URL: https://www.openmp.org/wp-content/uploads/OpenMP-API-Specification-5-1.pdf.

[OpenACC3.1] *The OpenACC Application Programming Interface v3.1.* The OpenACC Organization. Nov. 2020.

[Ouy+20] J. Ouyang et al. "Baidu Kunlun An AI processor for diversified workloads." In: *2020 IEEE Hot Chips 32 Symposium (HCS).* Los Alamitos, CA, USA: IEEE Computer Society, Aug. 2020, pp. 1–18. DOI: 10.1109/HCS49909.2020.9220641. URL: https://doi.ieeecomputersociety.org/10.1109/HCS49909.2020.9220641.

[Öze+18] G. Özen, S. Atzeni, M. Wolfe, A. Southwell, and G. Klimowicz. "OpenMP GPU Offload in Flang and LLVM." In: *2018 IEEE/ACM 5th Workshop on the LLVM Compiler Infrastructure in HPC (LLVM-HPC)*. Nov. 2018, pp. 1–9. DOI: 10.1109/LLVM-HPC.2018.8639434.

[Pat+15] Anuj Pathania, Alexandru Eugen Irimiea, Alok Prakash, and Tulika Mitra. "Power-Performance Modelling of Mobile Gaming Workloads on Heterogeneous MP-SoCs." In: *Proceedings of the 52nd Annual Design Automation Conference*. DAC '15. San Francisco, California: Association for Computing Machinery, 2015. ISBN: 9781450335201. DOI: 10.1145/2744769.2744894. URL: https://doi.org/10.1145/2744769.2744894.

[PCI09] PCI-SIG. *PCIe Address Translation Services (ATS)*. Standard Spec. Jan. 2009.

[PD10] S. Pasricha and N. Dutt. *On-Chip Communication Architectures: System on Chip Interconnect*. Elsevier Science, 2010. ISBN: 9780080558288.

[PHB14] B. Pichai, L. Hsu, and A. Bhattacharjee. "Architectural Support for Address Translation on GPUs: Designing Memory Management Units for CPU/GPUs with Unified Address Spaces." In: *Proceedings of the 19th International Conference on Architectural Support for Programming Languages and Operating Systems*. ASPLOS '14. Salt Lake City, Utah, USA: ACM, 2014, pp. 743–758. ISBN: 978-1-4503-2305-5. DOI: 10.1145/2541940.2541942. URL: http://doi.acm.org/10.1145/2541940.2541942.

[PHW14] J. Power, M. D. Hill, and D. A. Wood. "Supporting x86-64 address translation for 100s of GPU lanes." In: *2014 IEEE 20th International Symposium on High Performance Computer Architecture (HPCA)*. Feb. 2014, pp. 568–578. DOI: 10.1109/HPCA.2014.6835965.

[POSIX.1-2017] *POSIX.1-2017 (IEEE Std 1003.1-2017).* IEEE and The Open Group. 2018.

[Pow+15] J. Power, J. Hestness, M. S. Orr, M. D. Hill, and D. A. Wood. "gem5-gpu: A Heterogeneous CPU-GPU Simulator." In: *IEEE Computer Architecture Letters* 14.1 (2015), pp. 34–36. DOI: 10.1109/LCA.2014.2299539.

[PPT21] BM Prabhu Prasad, Khyamling Parane, and Basavaraj Talawar. "FPGA friendly NoC simulation acceleration framework employing the hard blocks." In: *Computing* (2021), pp. 1–23.

[Pul+12] Kari Pulli, Anatoly Baksheev, Kirill Kornyakov, and Victor Eruhimov. "Real-time computer vision with OpenCV." In: *Communications of the ACM* 55.6 (2012), pp. 61–69. URL: https://research.nvidia.com/sites/default/files/pubs/2012-06_Realtime-Computer-Vision/OpenCV_CACM_p61-pulli.pdf.

[Qua20] Qualcomm Inc. *Snapdragon 865 5G Mobile Platform.* 2020.

[Raj+13] Bharghava Rajaram, Vijay Nagarajan, Susmit Sarkar, and Marco Elver. "Fast RMWs for TSO: Semantics and Implementation." In: *PLDI.* 2013.

[Ran+21] Parthasarathy Ranganathan et al. "Warehouse-Scale Video Acceleration: Co-Design and Deployment in the Wild." In: *Proceedings of the 26th ACM International Conference on Architectural Support for Programming Languages and Operating Systems.* ASPLOS 2021. Virtual, USA: Association for Computing Machinery, 2021, pp. 600–615. ISBN: 9781450383172. DOI: 10.1145/3445814.3446723. URL: https://doi.org/10.1145/3445814.3446723.

[RAS10] I RAS. "GRAPHITE-OpenCL: Generate OpenCL code from parallel loops." In: *GCC Developers Summit. Citcseer* (2010), p. 9.

[Red+16] Joseph Redmon, Santosh Divvala, Ross Girshick, and Ali Farhadi. "You Only Look Once: Unified, Real-Time Object Detection." In: *2016 IEEE Conference on Computer Vision and Pattern Recognition (CVPR)*. 2016, pp. 779–788. DOI: 10.1109/CVPR.2016.91.

[Red+18] Jason Redgrave, Albert Meixner, Nathan Goulding-Hotta, Artem Vasilyev, and Ofer Shacham. "Pixel Visual Core: Google's Fully Programmable Image, Vision, and AI Processor for Mobile Devices." In: *2018 IEEE International Symposium on High Performance Chips (HOTCHIPS '30)*. 2018. URL: https://www.hotchips.org/hc30/1conf/1.02_Google_HC30.Google.JasonRedgrave.V01.pdf.

[Reu+20] A. Reuther, P. Michaleas, M. Jones, V. Gadepally, S. Samsi, and J. Kepner. "Survey of Machine Learning Accelerators." In: *IEEE High Performance Extreme Computing Conference (HPEC)*. 2020, pp. 1–12. DOI: 10.1109/HPEC43674.2020.9286149.

[Rey+12] Ruymán Reyes, Iván López-Rodriguez, Juan J. Fumero, and Francisco de Sande. "accULL: An OpenACC Implementation with CUDA and OpenCL Support." In: *Proceedings of the 2012 European Conference on Parallel Processing*. EuroPar '12. Springer, 2012, pp. 871–882. DOI: 10.1007/978-3-642-32820-6_86. URL: https://doi.org/10.1007/978-3-642-32820-6_86.

[Ros+14a] D. Rossi, I. Loi, G. Haugou, and L. Benini. "Ultra-low-latency Lightweight DMA for Tightly Coupled Multi-core Clusters." In: *Proceedings of the 11th ACM Conference on Computing Frontiers*. CF '14. Cagliari, Italy: ACM, 2014, 15:1–15:10. ISBN: 978-1-4503-2870-8. DOI: 10.1145/2597917.2597922. URL: http://doi.acm.org/10.1145/2597917.2597922.

[Ros+14b] Davide Rossi, Igor Loi, Francesco Conti, Giuseppe Tagliavini, Antonio Pullini, and Andrea Marongiu.

"Energy efficient parallel computing on the PULP platform with support for OpenMP." In: *Electrical & Electronics Engineers in Israel (IEEEI), 2014 IEEE 28th Convention of.* IEEE. 2014, pp. 1–5.

[Ros+17] D. Rossi et al. "Energy-Efficient Near-Threshold Parallel Computing: The PULPv2 Cluster." In: *IEEE Micro* 37.5 (Sept. 2017), pp. 20–31. ISSN: 0272-1732. DOI: 10.1109/MM.2017.3711645.

[Ros+21] Davide Rossi, Francesco Conti, Manuel Eggiman, Stefan Mach, Alfio Di Mauro, Marco Guermandi, Giuseppe Tagliavini, Antonio Pullini, Igor Loi, Jie Chen, Eric Flamand, and Luca Benini. "A 1.3TOPS/W @ 32GOPS Fully Integrated 10-Core SoC for IoT End-Nodes with 1.7uW Cognitive Wake-Up From MRAM-Based State-Retentive Sleep Mode." In: *2021 IEEE International Solid- State Circuits Conference (ISSCC).* Vol. 64. 2021, pp. 60–62. DOI: 10.1109/ISSCC42613.2021.9365939.

[RS99] A. Roth and G. S. Sohi. "Effective jump-pointer prefetching for linked data structures." In: *Proceedings of the 26th International Symposium on Computer Architecture (Cat. No.99CB36367).* 1999, pp. 111–121. DOI: 10.1109/ISCA.1999.765944.

[Sad+17] S. K. Sadasivam, B. W. Thompto, R. Kalla, and W. J. Starke. "IBM Power9 Processor Architecture." In: *IEEE Micro* (2017).

[SBH15] Hermann Schweizer, Maciej Besta, and Torsten Hoefler. "Evaluating the cost of atomic operations on modern architectures." In: *PACT.* 2015.

[SBL15] Erik Smistad, Mohammadmehdi Bozorgi, and Frank Lindseth. "FAST: Framework for heterogeneous medical image computing and visualization." In: *International Journal of Computer Assisted Radiology and Surgery* (Feb. 2015). DOI: 10.1007/s11548-015-1158-5.

[Sch+20] Fabian Schuiki, Andreas Kurth, Tobias Grosser, and Luca Benini. "LLHD: A Multi-Level Intermediate Representation for Hardware Description Languages." In: *Proceedings of the 41st ACM SIGPLAN Conference on Programming Language Design and Implementation.* PLDI 2020. London, UK: Association for Computing Machinery, 2020, pp. 258–271. ISBN: 9781450376136. DOI: 10.1145/3385412.3386024. URL: https://doi.org/10.1145/3385412.3386024.

[SDS00] A. Saulsbury, F. Dahlgren, and P. Stenström. "Recency-based TLB Preloading." In: *Proceedings of the 27th Annual International Symposium on Computer Architecture.* ISCA '00. Vancouver, British Columbia, Canada: ACM, 2000. ISBN: 1-58113-232-8. DOI: 10.1145/339647.339666. URL: http://doi.acm.org/10.1145/339647.339666.

[SGG18] A. Silberschatz, P.B. Galvin, and G. Gagne. *Operating system concepts.* eng. 10th edition. Hoboken, N.J: Wiley, 2018. ISBN: 9781119299677.

[SGS10] John E Stone, David Gohara, and Guochun Shi. "OpenCL: A parallel programming standard for heterogeneous computing systems." In: *Computing in science & engineering* 12.3 (2010), p. 66.

[Sha98] T. Shanley. In: *Pentium Pro and Pentium II System Architecture.* Mindshare PC System Architecture Series. Addison-Wesley, 1998. Chap. 22 Paging Enhancements. ISBN: 9780201309737.

[She+03] T. Sherwood, E. Perelman, G. Hamerly, S. Sair, and B. Calder. "Discovering and exploiting program phases." In: *IEEE Micro* 23.6 (Nov. 2003), pp. 84–93. ISSN: 0272-1732. DOI: 10.1109/MM.2003.1261391.

[Sie21] *Emulation – A Job Management Strategy to Maximize Use.* White Paper. Siemens, 2021. URL: https://resources.sw.siemens.com/en-US/white-paper-emulation-a-job-management-strategy-to-maximize-use.

[SiF18] SiFive Inc. *TileLink Specification.* 2018.

[SiF19] *SiFive TileLink Specification v1.8.0.* SiFive Inc. 2019.

[Sil10] *Wishbone B4 SoC Interconnection Architecture.* Silicore Corp. 2010.

[Sin+13] Amit Kumar Singh, Muhammad Shafique, Akash Kumar, and Jörg Henkel. "Mapping on multi/many-core systems: Survey of current and emerging trends." In: *2013 50th ACM/EDAC/IEEE Design Automation Conference (DAC).* 2013, pp. 1–10. DOI: 10.1145/2463209.2488734.

[SLB10] A. Strano, D. Ludovici, and D. Bertozzi. "A library of dual-clock FIFOs for cost-effective and flexible MPSoC design." In: *IEEE SAMOS.* 2010.

[Smi20] Ryan Smith. "NVIDIA Ampere Unleashed: NVIDIA Announces New GPU Architecture, A100 GPU, and Accelerator." In: *AnandTech* (2020).

[Son+09] S. W. Son, M. Kandemir, M. Karakoy, and D. Chakrabarti. "A Compiler-directed Data Prefetching Scheme for Chip Multiprocessors." In: *Proceedings of the 14th ACM SIGPLAN Symposium on Principles and Practice of Parallel Programming.* PPoPP '09. Raleigh, NC, USA: ACM, 2009, pp. 209–218. ISBN: 978-1-60558-397-6. DOI: 10.1145/1504176.1504208. URL: http://doi.acm.org/10.1145/1504176.1504208.

[Sto+13] Eric Stotzer, Ajay Jayaraj, Murtaza Ali, Arnon Friedmann, Gaurav Mitra, Alistair P. Rendell, and Ian Lintault. "OpenMP on the Low-Power TI Keystone II ARM/DSP System-on-Chip." In: *OpenMP in the Era of Low Power Devices and Accelerators.* Ed. by Alistair P. Rendell, Barbara M. Chapman, and Matthias S. Müller. Berlin, Heidelberg: Springer Berlin Heidelberg, 2013, pp. 114–127. ISBN: 978-3-642-40698-0.

[Stu+15] J. Stuecheli, B. Blaner, C. R. Johns, and M. S. Siegel. "CAPI: A Coherent Accelerator Processor Interface." In: *IBM Journal of R&D* 59.1 (Jan. 2015), 7:1–7:7. ISSN: 0018-8646. DOI: 10.1147/JRD.2014.2380198.

[Syn21] *ZeBu EP1: Industry's Fastest Billion Gates Emulation System.* Synopsys. 2021. URL: https://www.synopsys.com/verification/emulation/zebu-ep1.html.

[Sze+17] Vivienne Sze, Yu-Hsin Chen, Tien-Ju Yang, and Joel S. Emer. "Efficient Processing of Deep Neural Networks: A Tutorial and Survey." In: *Proceedings of the IEEE* 105.12 (2017), pp. 2295–2329. DOI: 10.1109/JPROC.2017.2761740.

[Tan+10a] Z. Tan et al. "A Case for FAME: FPGA Architecture Model Execution." In: *Proceedings of the 37th Annual International Symposium on Computer Architecture.* ISCA '10. Saint-Malo, France: ACM, 2010, pp. 290–301. ISBN: 978-1-4503-0053-7. DOI: 10.1145/1815961.1815999. URL: http://doi.acm.org/10.1145/1815961.1815999.

[Tan+10b] Z. Tan et al. "RAMP Gold: An FPGA-based Architecture Simulator for Multiprocessors." In: *Proceedings of the 47th Design Automation Conference.* DAC '10. Anaheim, California: ACM, 2010, pp. 463–468. ISBN: 978-1-4503-0002-5. DOI: 10.1145/1837274.1837390. URL: http://doi.acm.org/10.1145/1837274.1837390.

[Ter19] E. Terry. "Silicon at the Heart of HoloLens 2." In: *2019 IEEE Hot Chips 31 Symposium (HCS).* Los Alamitos, CA, USA: IEEE Computer Society, Aug. 2019, pp. 1–26. DOI: 10.1109/HOTCHIPS.2019.8875669. URL: https://doi.ieeecomputersociety.org/10.1109/HOTCHIPS.2019.8875669.

[Tex19a] Texas Instruments Inc. *66AK2Hxx Multicore DSP+ARM KeyStone II System-on-Chip (SoC) datasheet (Rev.*

G). Oct. 2019. URL: http://www.ti.com/lit/ds/symlink/66ak2h14.pdf.

[Tex19b] Texas Instruments Inc. *TDA2x ADAS Applications Processor 17mm Package (AAS) Silicon Revision 2.0 datasheet (Rev. F)*. June 2019. URL: http://www.ti.com/lit/ds/sprs952f/sprs952f.pdf.

[Tex19c] Texas Instruments Inc. *TMS320C6678 Multicore Fixed and Floating-Point Digital Signal Processor datasheet (Rev. E)*. Mar. 2019. URL: http://www.ti.com/lit/ds/symlink/tms320c6678.pdf.

[Tin+20] Blaise Tine, Fares Elsabbagh, Lee Seyong, Jeff Vetter, and Hyesoon Kim. "Cash: A Single-Source Hardware-Software Codesign Framework for Rapid Prototyping." In: *Proceedings of the 2020 ACM/SIGDA International Symposium on Field-Programmable Gate Arrays*. FPGA '20. Seaside, CA, USA: Association for Computing Machinery, 2020, p. 321. ISBN: 9781450370998. DOI: 10.1145/3373087.3375340. URL: https://doi.org/10.1145/3373087.3375340.

[Tra+16] A. Traber et al. *PULPino: A small single-core RISC-V SoC*. Jan. 2016. URL: https://riscv.org/wp-content/uploads/2016/01/Wed1315-PULP-riscv3_noanim.pdf.

[Tre18] Andreas Tretter. "On Efficient Data Exchange in Multicore Architectures." en. PhD thesis. Zurich: ETH Zurich, 2018. DOI: 10.3929/ethz-b-000309314.

[Tze92] N.-F. Tzeng. "A cost-effective combining structure for large-scale shared-memory multiprocessors." In: *IEEE TC* (1992).

[Uba+12] R. Ubal, B. Jang, P. Mistry, D. Schaa, and D. Kaeli. "Multi2Sim: A simulation framework for CPU-GPU computing." In: *21st IEEE International Conference on Parallel Architectures and Compilation Techniques (PACT)*. 2012, pp. 335–344.

[Vaz+18] Sudharshan S. Vazhkudai et al. "The Design, Deployment, and Evaluation of the CORAL Pre-exascale Systems." In: *Proceedings of the International Conference for High Performance Computing, Networking, Storage, and Analysis.* SC '18. Dallas, Texas: IEEE Press, 2018, 52:1–52:12. DOI: 10.1109/SC.2018.00055. URL: https://doi.org/10.1109/SC.2018.00055.

[VBT19] A. Venkat, H. Basavaraj, and D. M. Tullsen. "Composite-ISA Cores: Enabling Multi-ISA Heterogeneity Using a Single ISA." In: *2019 IEEE International Symposium on High Performance Computer Architecture (HPCA).* Feb. 2019, pp. 42–55. DOI: 10.1109/HPCA.2019.00026.

[Ves+16] J. Vesely, A. Basu, M. Oskin, G. H. Loh, and A. Bhattacharjee. "Observations and opportunities in architecting shared virtual memory for heterogeneous systems." In: *2016 IEEE International Symposium on Performance Analysis of Systems and Software (ISPASS).* Apr. 2016. DOI: 10.1109/ISPASS.2016.7482091.

[VMB17] Pirmin Vogel, Andrea Marongiu, and Luca Benini. "Lightweight Virtual Memory Support for Zero-Copy Sharing of Pointer-Rich Data Structures in Heterogeneous Embedded SoCs." In: *IEEE Transactions on Parallel and Distributed Systems* 28.7 (2017), pp. 1947–1959. DOI: 10.1109/TPDS.2016.2645219.

[VMB18] Pirmin Vogel, Andrea Marongiu, and Luca Benini. "Exploring shared virtual memory for FPGA accelerators with a configurable IOMMU." In: *IEEE Transactions on Computers* (2018).

[Vog+15] P. Vogel et al. "Lightweight Virtual Memory Support for Many-core Accelerators in Heterogeneous Embedded SoCs." In: CODES '15 (2015), pp. 45–54. URL: http://dl.acm.org/citation.cfm?id=2830840.2830846.

[Vog+17] Pirmin Vogel, Andreas Kurth, Johannes Weinbuch, Andrea Marongiu, and Luca Benini. "Efficient Virtual Memory Sharing via On-Accelerator Page Table Walking in Heterogeneous Embedded SoCs." In: *ACM Transactions on Embedded Computing Systems* 16.5s (Sept. 2017). ISSN: 1539-9087. DOI: 10.1145/3126560. URL: https://doi.org/10.1145/3126560.

[VT14] Ashish Venkat and Dean M. Tullsen. "Harnessing ISA Diversity: Design of a heterogeneous-ISA Chip Multiprocessor." In: *Proceeding of the 41st Annual International Symposium on Computer Architecuture*. ISCA '14. Minneapolis, Minnesota, USA: IEEE Press, 2014, pp. 121–132. ISBN: 978-1-4799-4394-4. URL: http://dl.acm.org/citation.cfm?id=2665671.2665692.

[Wan+10] Q. Wang et al. "An FPGA Based Hybrid Processor Emulation Platform." In: *FPL*. Aug. 2010, pp. 25–30. DOI: 10.1109/FPL.2010.16.

[Wan+14] Danyao Wang, Charles Lo, Jasmina Vasiljevic, Natalie Enright Jerger, and J. Gregory Steffan. "DART: A Programmable Architecture for NoC Simulation on FPGAs." In: *IEEE Transactions on Computers* 63.3 (2014), pp. 664–678. DOI: 10.1109/TC.2012.121.

[Wat+11] Andrew Waterman, Yunsup Lee, David A Patterson, and Krste Asanovic. *The RISC-V instruction set manual.* 2011.

[Wat+19] Andrew Waterman, Yunsup Lee, Rimas Avizienis, David A Patterson, and Krste Asanović. *The RISC-V Instruction Set Manual, Volume II: Privileged Architecture.* Version 20190608-Priv-MSU-Ratified. 2019.

[Wat16] A. Waterman. *Design of the RISC-V Instruction Set Architecture.* Jan. 2016. URL: http://www2.eecs.berkeley.edu/Pubs/TechRpts/2016/EECS-2016-1.html.

[Whe19] Bob Wheeler. "Tomahawk 4 Switch First to 25.6 Tbps." In: *Microprocessor Report* (2019).

[Wol+17] M. Wolfe, S. Lee, J. Kim, X. Tian, R. Xu, S. Chandrasekaran, and B. Chapman. "Implementing the OpenACC Data Model." In: *2017 IEEE International Parallel and Distributed Processing Symposium Workshops (IPDPSW)*. May 2017, pp. 662–672. DOI: 10.1109/IPDPSW.2017.85.

[Woo+14] Jonathan Woodruff, Robert N.M. Watson, David Chisnall, Simon W. Moore, Jonathan Anderson, Brooks Davis, Ben Laurie, Peter G. Neumann, Robert Norton, and Michael Roe. "The CHERI Capability Model: Revisiting RISC in an Age of Risk." In: *Proceeding of the 41st Annual International Symposium on Computer Architecuture*. ISCA '14. Minneapolis, Minnesota, USA: IEEE Press, 2014, pp. 457–468. ISBN: 978-1-4799-4394-4. URL: http://dl.acm.org/citation.cfm?id=2665671.2665740.

[Xil16] *Zynq-7000 All Programmable SoC Overview*. Xilinx Inc. 2016.

[Xil17] Xilinx Inc. *Zynq UltraScale+ MPSoC Data Sheet: Overview*. Advance Product Specification. Feb. 2017.

[Yan+10] Canqun Yang, Feng Wang, Yunfei Du, Juan Chen, Jie Liu, Huizhan Yi, and Kai Lu. "Adaptive Optimization for Petascale Heterogeneous CPU/GPU Computing." In: *2010 IEEE International Conference on Cluster Computing*. 2010, pp. 19–28. DOI: 10.1109/CLUSTER.2010.12.

[YM16] Chaoran Yang and John Mellor-Crummey. "A Wait-free Queue As Fast As Fetch-and-add." In: *PPoPP*. 2016.

[Zah17] Mohamed Zahran. "Heterogeneous Computing: Here to Stay." In: *Commun. ACM* 60.3 (Feb. 2017), pp. 42–45. ISSN: 0001-0782. DOI: 10.1145/3024918. URL: https://doi.org/10.1145/3024918.

[Zah19] Mohamed Zahran. *Heterogeneous Computing: Hardware and Software Perspectives.* New York, NY, USA: Association for Computing Machinery, 2019. ISBN: 9781450360975.

[Zar+19] F. Zaruba, F. Schuiki, S. Mach, and L. Benini. "The Floating Point Trinity: A Multi-modal Approach to Extreme Energy-Efficiency and Performance." In: *IEEE ICECS.* 2019.

[Zar+20] Florian Zaruba, Fabian Schuiki, Torsten Hoefler, and Luca Benini. "Snitch: A tiny Pseudo Dual-Issue Processor for Area and Energy Efficient Execution of Floating-Point Intensive Workloads." In: *IEEE Transactions on Computers* (2020), pp. 1–1. DOI: 10.1109/TC.2020.3027900.

[ZB19] Florian Zaruba and Luca Benini. "The Cost of Application-Class Processing: Energy and Performance Analysis of a Linux-Ready 1.7-GHz 64-Bit RISC-V Core in 22-nm FDSOI Technology." In: *IEEE Transactions on Very Large Scale Integration (VLSI) Systems* 27.11 (2019), pp. 2629–2640. DOI: 10.1109/TVLSI.2019.2926114.

[Zha+20] Jialiang Zhang, Yue Zha, Nicholas Beckwith, Bangya Liu, and Jing Li. "MEG: A RISCV-Based System Emulation Infrastructure for Near-Data Processing Using FPGAs and High-Bandwidth Memory." In: *ACM Trans. Reconfigurable Technol. Syst.* 13.4 (Sept. 2020). ISSN: 1936-7406. DOI: 10.1145/3409114. URL: https://doi.org/10.1145/3409114.

[ZHA19] Fatma Zahran, Amal O. Hamada, and Mohamed Azab. "Cooperative Heterogeneous IoT for Health." In: *2019 IEEE 9th Symposium on Computer Applications & Industrial Electronics (ISCAIE).* 2019, pp. 352–357. DOI: 10.1109/ISCAIE.2019.8743973.

[Zim+16] B. Zimmer et al. "A RISC-V Vector Processor With Simultaneous-Switching Switched-Capacitor DC-DC Converters in 28 nm FDSOI." In: *IEEE JSSC* 51.4

(Apr. 2016), pp. 930–942. ISSN: 0018-9200. DOI: 10.1109/JSSC.2016.2519386.

[ZSB20] F. Zaruba, F. Schuiki, and L. Benini. “A 4096-core RISC-V Chiplet Architecture for Ultra-efficient Floating-point Computing.” In: *2020 IEEE Hot Chips 32 Symposium (HCS)*. Los Alamitos, CA, USA: IEEE Computer Society, Aug. 2020, pp. 1–24. DOI: 10.1109/HCS49909.2020.9220474. URL: https://doi.ieeecomputersociety.org/10.1109/HCS49909.2020.9220474.

[Zuc+21] Joseph Zuckerman, Davide Giri, Jihye Kwon, Paolo Mantovani, and Luca P. Carloni. “Cohmeleon: Learning-Based Orchestration of Accelerator Coherence in Heterogeneous SoCs.” In: *MICRO-54: 54th Annual IEEE/ACM International Symposium on Microarchitecture*. MICRO ’21. Virtual Event, Greece: Association for Computing Machinery, 2021, pp. 350–365. ISBN: 9781450385572. DOI: 10.1145/3466752.3480065. URL: https://doi.org/10.1145/3466752.3480065.

Appendix C

Curriculum Vitae

Andreas Kurth was born in Zürich, Switzerland, in 1990. He received his MSc degree *with distinction* in Electrical Engineering and Information Technology from ETH Zurich in 2017. He then joined the Digital Circuits and Systems Group at the Integrated Systems Laboratory (IIS) at ETH Zurich and received his PhD degree under the supervision of Prof. Dr. Luca Benini in 2022. His research interests include the architecture and programming of heterogeneous systems on chip and accelerator-rich computing systems. He has received the ESTIMedia Best Paper Award in 2016 and was nominated for the ETH Medal in 2017 and the DAC Best Paper Award in 2020.

SERIES IN MICROELECTRONICS

#167: Thomas von Büren, Body-Worn Inertial Electromagnetic Micro-Generators. ISBN 3-86628-064-5

#168: Frank K. Gürkaynak, GALS System Design: Side Channel Attack Secure Cryptographic Accelerators. ISBN 3-86628-065-3

#169: Andreas Burg, VLSI Circuits for MIMO Communication Systems. ISBN 3-86628-066-1

#170: Melanie Etherton, Charged Device Model (CDM) ESD in ICs: Physics, Modeling, and Circuit Simulation. ISBN 3-86628-067-X

171: Christian Kromer, 10 Gb/s to 40 Gb/s Receiver for High-Density Optical Interconnects in 80-nm CMOS. ISBN 3-86628-069-6

#172: Corrado Carta, BiCMOS Radio-Frequency Front-Ends for Wireless-LAN Receivers. ISBN 3-86628-071-8

#173: Timm Höhr, Quantum-Mechanical Modeling of Transport Parameters for MOS-Devices. ISBN 3-86628-087-4

#174: Ilian Kouchev, Design of a Highly Linear Direct-Conversion Receiver for Third-Generation Mobile Communications. ISBN 3-86628-093-9

#175: Chiara Martelli, Multi-Standard Low-Power Base-Band Digital Receiver, Enhanced for HSDPA. ISBN 3-86628-094-7

#176: Walter Oesch, Controlling Software for EMF Laboratory Studies. ISBN 3-86628-097-1

#177: Stefan Odermatt, Physics and Simulation of Semiconductor Lasers: Static and Dynamic Characteristics. ISBN 3-86628-111-0

#178: José Miguel Ruiz Palmero, Physical InP-Based HBT Models for Ultimate Digital Circuit Optimization. ISBN 3-86628-117-X

#179: Lutz Schneider, Multidimensional Modeling and Simulation of Wavelength-Tunable Semiconductor Lasers. ISBN 3-86628-122-6

#180: Marco Buzzo, Dopant Imaging and Profiling of Wide Bandgap Semiconductor Devices. ISBN 3-86628-124-2

#181: Beat Sahli, Ab Initio Molecular Dynamics Simulation of Diffusion in Silicon. ISBN 3-86628-133-1

#182: Ulrich Glaser, Complex ESD Protection Elements and Issues in Decananometre CMOS Technologies. ISBN 3-86628-135-8

#183: Chiara Corvasce, Mobility and Impact Ionization in Silicon at High Temperature. ISBN 3-86628-136-6

#184: Stefan Benkler, Robust Conformal Subcell Modeling for Electromagnetic Simulations in Time Domain. ISBN 3-86628-137-4

#185: Gion Sialm, VCSEL Modeling and CMOS Trans-mitters up to 40 Gb/s for High-Density Optical Links. ISBN 3-86628-143-9

#186: Mathieu Luisier, Quantum Transport Beyond the Effective Mass Approximation. ISBN 3-86628-149-8

#187: Flavio Carbognani, Low-Power Techniques for Low-Frequency VLSI Applications. ISBN 3-86628-161-7

#188: Marc Simon Wegmüller, Intra-Body Communication for Biomedical Sensor Networks. ISBN 3-86628-162-5

189: Valerio Laino, Performance Analysis of Edge Emitting Lasers in the Mid Infra-Red and Visible Spectrum. ISBN 3-86628-169-2

190: Christoph David Perels, Frame-Based MIMO-OFDM Systems: Impairment Estimation and Compensation. ISBN 3-86628-172-2

191: Verónica J. Berdinas Torres, Exposure Systems and Dosimetry of Large-Scale In Vivo Studies. ISBN 3-86628-180-3

192: Simon Häne, VLSI Circuits for MIMO-OFDM Physical Layer. ISBN 3-86628-196-X

193: Felix Bürgin, Low-Power Circuit Architectures and Clocking Strategies for Digital Hearing Aids. ISBN 3-86628-211-7

194: Luca Sponton, From manufacturing variability to process-aware circuit simulation. 3-86628-230-3

195: Yves Saad, TCAD-Based Three-Dimensional Modeling of Non-Volatile Memories. ISBN 3-86628-237-0

196: Stefan Schild, Advanced Material Modeling in EM-FDTD ISBN 3-86628-238-9

197: Sebastian Steiger, Modelling Nano-LEDs. ISBN 3-86628-257-5

198: Alexandra Baecker, A TCAD Analysis of Long-Wavelength Vertical-Cavity Surface-Emitting Lasers. ISBN 3-86628-258-3

199: Martin Loeser, Theory and Design of Broadband Active Opto-electronic Devices. ISBN 3-86628-276-1

200: Stefan Eberli, Application-Specific Processor for MIMO-OFDM Software-Defined Radio. ISBN 3-86628-285-0

201: Ratko G. Veprek, omputational Modeling of Semiconductor Nanostructures for Optoelectronics. ISBN 3-86628-286-9

202: Christoph Studer, Iterative MIMO Decoding: Algorithms and VLSI Implementation Aspects. ISBN 3-86628-288-5

203: Peter Jan Lüthi, VLSI Circuits for MIMO Preprocessing. ISBN 3-86628-295-8

204: Urs Hammer, Sub-micron InP/GaAsSb/InP Double Hetero-junction Bipolar Transistors for Ultra High-Speed Digital Integrated Circuits. ISBN 3-86628-301-6

205: Sven Kühn, EMF Risk Assessment: Exposure Assessment and Compliance Testing in Complex Environments. ISBN 3-86628-309-1

206: Jürg Treichler, Preserving High Resolution in Deep-Submicron CMOS Pipelined A/D. ISBN 3-86628-316-4

207: Philipp Kreuter, Modeling of Electrically Pumped Vertical-External-Cavity Surface-Emitting Lasers. ISBN 978-3-86628-317-6

208: Thomas Christen, Multi-Mode Delta-Sigma A/D-Converters for Multi-Standard Wireless Receivers. ISBN 978-3-86628-318-3

209: Christian Benkeser, Power Efficiency and the Mapping of Communication Algorithms into VLSI. ISBN 978-3-86628-323-7

210: Sven Ebert, EMF Risk Assessment: Exposure Systems for Large-Scale Laboratory and Experimental Provocation Studies. ISBN 978-3-86628-331-2

211: Martin Frey, Scattering in Nanoscale Devices. ISBN 978-3-86628-353-4

212: Kilian Vollenweider, Dopant Clustering and Diffusion in Silicon. ISBN 978-3-86628-354-1

213: Markus Wenk, MIMO-OFDM Testbed: Challenges, Implementations, and Measurement Results. ISBN 978-3-86628-366-4

214: Luca Henzen, VLSI Circuits for Cryptographic Authentication. ISBN 978-3-86628-367-1

215: Aniello Esposito, Band Structure Effects and Quantum Transport. ISBN 978-3-86628-378-7

216: Hektor Meier, Design, Characterization and Simulation of Avalanche Photodiodes. ISBN 978-3-86628-380-0

217: Denis Dolgos, Full-Band Monte Carlo Simulation of Single Photon Avalanche Diodes. ISBN 978-3-86628-421-0

218: Neviana Nikoloski, EMF Risk Assessment: Exposure Assessment and Specialized Exposure Setups. ISBN 978-3-86628-422-7

219: Patrick Mächler, VLSI Architectures for Compressive Sensing and Sparse Signal Recovery. ISBN 978-3-86628-446-3

220: Marcus Deppner, Design of Nanorod-LEDs using Computational Modelling. ISBN 978-3-86628-459-3

221: Pierre Greisen, Hardware Architectures for Real-Time Video Processing and View Synthesis. ISBN 978-3-86628-461-6

222: Vincent Peikert, Utilizing Wavelets to Solve High-Dimensional Transport Equations in Nano-Devices. ISBN 978-3-86628-466-1

223: Cédric Bessire, Semiconducting Nanowire Tunnel Devices. ISBN 978-3-86628-471-3

224: Artur Scheinemann, Modelling of Leakage Currents Induced by Extended Defects in Extra-Functionality Devices. ISBN 978-3-86628-504-0

225: Zhelio Andreev, Physics-based Simulation of III-V Nitride LEDs: The Role of Polarization. ISBN 978-3-86628-542-2

226: Reto Rhyner, Quantum Transport Beyond The Ballistic Limit. ISBN 978-3-86628-545-3

227: Philipp Mensch, Thermoelectric Characterization of InAs Nanowires. ISBN 978-3-86628-546-0

228: Aron Szabo, Dissipative quantum transport simulations in two-dimensional semiconductor devices from first principles. ISBN 978-3-86628-570-5

229: Harald Kröll, An Evolved EDGE System-on-Chip for the Cellular Internet of Things. ISBN 978-3-86628-572-9

230: Philipp Christoph Schönle, A Power Efficient Spectrophotometry & PPG Integrated Circuit for Mobile Medical Instruments. ISBN 978-3-86628-590-3

231: Michael Andreas Gautschi, Design of Energy-Efficient Processing Elements for Near-Threshold Parallel Computing.ISBN 978-3-86628-595-8

232: Benjamin Sporrer, Integrated Broadband Receivers for Magnetic. Resonance Imaging. ISBN 978-3-86628-607-8

233: David Emanuel Bellasi, Toward Energy-Proportional Compressive Sensors. ISBN 978-3-86628-618-4

234: Pirmin Robert Vogel, Shared Virtual Memory for Heterogeneous Embedded Systems on Chips. ISBN 978-3-86628-623-8

235: Michael Schaffner, Energy-Efficient VLSI Architectures for Real-Time and 3D Video Processing ISBN 978-3-86628-624-5

236: Sandro Steffano Belfanti, Design and Implementation of Efficient VLSI Solutions for 3G and 4G Mobile Communications. ISBN 978-3-86628-635-1

237: Pascal Alexander Hager, Design of Fully-Digital Medical Ultrasound Imaging Systems. ISBN 978-3-86628-649-8

238: Lukas Cavigelli, Towards Energy-Efficient Convolutional Neural Network Inference. ISBN 978-3-86628-651-1

239: Florian Michael Scheidegger, The concept of transprecision computing. ISBN 978-3-86628-689-4

240: Renzo Andri, Machine Learning Acceleration for Tightly Energy-Constrained Devices. ISBN 978-3-86628-693-1

241: Stefan Mach, Floating-Point Architectures for Energy-Efficient Transprecision Computing. ISBN 978-3-86628-719-8

242: Florian Zaruba, Energy-Efficient High-Performance Computing. ISBN 978-3-86628-720-4

243: Fabian Schuiki, Streaming Architectures for Extreme Energy Efficiency in High-Performance Computing. ISBN 978-3-86628-725-9

Hartung-Gorre Verlag, Konstanz
http://www.hartung-gorre.de